THE FUTURE OF AEROSPACE

Proceedings of a Symposium

Held in Honor of

Alexander H. Flax

Home Secretary

National Academy of Engineering

NATIONAL ACADEMY PRESS
Washington, D.C. 1993

The National Academy of Engineering was established in 1964, under the charter of the National Academy of Sciences, as a parallel organization of outstanding engineers. It is autonomous in its administration and in the selection of its members, sharing with the National Academy of Sciences the responsibility for advising the federal government. The National Academy of Engineering also sponsors engineering programs aimed at meeting national needs, encourages education and research, and recognizes the superior achievement of engineers. Dr. Robert M. White is president of the National Academy of Engineering.

This volume consists of papers and speakers' remarks presented during a commemorative symposium in honor of outgoing NAE Home Secretary Alexander H. Flax. The symposium, entitled "The Future of Aerospace," was held 28 February 1992. The interpretations and conclusions expressed in the symposium papers are those of the authors and are not presented as the views of the council, officers, or staff of the National Academy of Engineering.

Library of Congress Catalog Card Number 93-83945

International Standard Book Number 0-309-04881-8

Additional copies of this report are available for sale from:

National Academy Press
2101 Constitution Avenue, N.W.
Washington, D.C. 20418

B-137

Printed in the United States of America

Contents

Preface

Few technological advances have affected the lives and dreams of individuals and the operations of companies and governments as much as the continuing development of flight. In the United States alone there are, on average, more than one million commercial enplanements a day, and air freight, once a "special express" service, is now the common form of transport for many goods. Air power and the battlefield control offered by airborne equipment and personnel are increasingly decisive elements of military conflicts, and outer space is still a boundless frontier for human and scientific exploration. And the technological frontiers of what is possible and what is likely continue to recede in front of us even as we advance. From space exploration to package transport, from military transport to passenger helicopter use, from passenger jumbo jets to tilt-rotor commuter planes, the future of flying is still rapidly developing.

The essays in this volume survey the state of progress and assess prospects for the future along several fronts of this constantly evolving frontier. More than that, they commemorate the life's work of Alexander H. Flax, who has contributed to the technological developments in many fields of aerospace engineering and to the institutions that guide those developments. For eight years as home secretary to the National Academy of Engineering, Al Flax also guided the Academy's membership affairs, helped oversee the Academy's diverse study program, and was a steady source of sound advice. It was a pleasure to work with him. So many of us have prized his friendship over many decades.

The five essays here are adapted from remarks prepared by their authors for a symposium held in Al Flax's honor at the Arnold and Mabel Beckman Center of the National Academies of Sciences and Engineering on 28 February 1992. Thus, it is important to acknowledge first the work of those authors, all members of the National Academy of Engineering, for making this volume possible and for contributing to our better understanding of aerospace issues. The participants in this symposium included many knowledgeable and leading figures in the areas of Al Flax's career. They contributed greatly to the formal discussion of the papers and in the informal occasions afforded by the symposium.

There are a few others to whom thanks are also due. Melvin Gipson and Maribeth Keitz, on the staff of the NAE Program Office, led in organizing the symposium. Dale Langford and Bette Janson prepared the volume for publication. Bruce Guile, director of the Program Office, provided direction and oversight for the project.

H. Guyford Stever
Former NAE Foreign Secretary
Symposium Chairman

Alexander H. Flax
Highlights of an Engineering Career

Born in Brooklyn, Alexander H. Flax received his early education there and attended New York University, where he received a B.S. in aeronautical engineering in 1940.

Flax began his professional career in 1940 as a stress analyst with the Curtiss-Wright Corporation, where within two years he became chief of the flutter and vibration group. He worked primarily on structural and dynamics problems and on methods for design analysis and testing. With the advent of electric strain gages, new opportunities arose to validate analytical methods, many of which were just emerging as replacements for the more empirical and judgmental methods of the past. Flax was very active in developing and applying these new approaches to the many aircraft types under development at Curtiss-Wright. These included the O-52—a high-wing, strut-braced observation aircraft and last of a line and an era; the P-40 fighter—the most advanced fighter available in quantity at the outbreak of World War II; the XP-60 and XP-62—experimental fighter aircraft never produced in quantity; the SB2C-1—a Navy dive bomber that entered service in 1943; and the C-46—a military transport aircraft extensively used to "fly the hump" in the China-Burma-India theater. Flax's work on these aircraft included stress analysis, flutter and vibration analysis, and advanced flight loads analysis.

In 1944 he moved to the Piasecki Helicopter Corporation (which later became the Vertol Division of the Boeing Company) as head of aerodynamics, structures, and weights—a position that in some larger companies was occupied by the chief technical engineer. At Piasecki, Flax was one of a small group of engineers who developed the world's first twin-rotor, tandem helicopter (known as the Navy XHRP-1 or, more informally, as "The Dogship"). Helicopter technology was then in its infancy even for single-rotor machines, and the additional complication of tandem rotors required starting from scratch on many questions of design, analysis, and testing. Nevertheless, Piasecki won two major design competitions—the HUP-1 for the Navy and HU-16 for the Army/Air Force. Descendants of the twin-tandem HUP-1 are still in service as the CH-46 and CH-47 transport helicopters seen most recently in the Persian Gulf conflict.

In 1946 Flax joined the Cornell Aeronautical Laboratory (now known as CALSPAN) as assistant head of the aeromechanics department. Continuing to do research on helicopter rotors, he and his colleagues built and flew, for purposes of research on rotor blade dynamics, what may have been the world's first flight-worthy and flight-demonstrated fiberglass composite rotor blades. It was perhaps twenty years after this work that composite fiber blades appeared in operational helicopters.

Soon after joining the Cornell Aeronautical Laboratory, Flax branched out into supersonic vehicle research, including supersonic aerodynamics, flight control, and ramjet propulsion. His research in wing theory and wing-body interferences in this period was widely recognized. He conceived of the perforated-wall wind tunnel, one of two wind-tunnel designs currently in use for testing in transonic flows both above and below the speed of sound. He was also one of the inventors of the wave superheater for generating "clean" airflows of several-second duration at temperatures previously attained only in rocket exhaust flows.

During his final years at Cornell Aeronautical Laboratory, as vice president and technical director, Flax exercised managerial and technical guidance over a wide variety of projects, many not particularly aeronautical. These included early work on neural network computers; early work on automotive crash safety, including seatbelts, energy-absorbing structures, and door locks; pioneering work on stability, control, and handling qualities of automobiles, including the statics and dynamics of tires; and early work on doppler radars for weather sensing. He received a Ph.D. in physics from the University of Buffalo in 1958.

Flax served as chief scientist of the U.S. Air Force from 1959 to 1961 and in 1963 was appointed assistant secretary of the Air Force for research and development. From that position he championed advanced aircraft engine development as an essential element of progress in both military and civilian aeronautics. In the 1960s he emphasized the Lightweight Engine Gas Generator Program and the Advanced Turbine Engine Gas Generator Program, work that reached fruition in the 1970s with the engines that went into the F-15 and F-16 fighter aircraft and the new generation of high-bypass engines for military and civil large, long-range transports. Another area of special attention was materials, especially the then-emerging field of high-strength, high-stiffness, lightweight fiber composites. This work, which began with boron fibers in the early 1960s and quickly expanded to include graphite fibers and Kevlar, has found particular application in helicopters and vertical takeoff aircraft, in some of which more than 40 percent of the structural weight may be in fiber composites.

Other areas of research emphasized during Flax's service with the Air Force included the development of precision-guided weapons and the corresponding aircraft targeting systems and their sensors. As a result of these research thrusts, laser and electro-optical guided bombs were quickly developed. Optical, infrared, and high-resolution radar sensor systems were vigorously pursued, and on-board computer capabilities were added as standard equipment or modular additions to all future fighter-attack aircraft. The result of all of these initiatives and their further exploitation became apparent in the Persian Gulf conflict.

Military space systems also fell within Flax's responsibilities in the Air Force. During his tenure the Defense Support Program, a satellite-borne infrared sensor for ballistic missile launch detection, which is still operational, reached full engineering development status, and the Titan III launch vehicle family reached operational flight status. Early versions of defense communications satellites including DSCS-1, TACSATCOM, and several Lincoln Experimental Satellites were developed and flown. The conceptual groundwork and applications studies for the Global Positioning System, now reaching full operational status, were laid down at that time. From 1965 to 1969 Flax held the additional position of director of the National Reconnaisance Office and was in charge of R&D, procurement, and operations of satellite systems for intelligence and military support that later also served as the primary "national technical means" for verification of arms control treaties.

In March 1969, Flax joined the Institute for Defense Analyses (IDA) as vice president of research and later that year became president; he retired in 1983. While at IDA, Flax oversaw activities in which support was provided to the Office of the Secretary of Defense and the Joint Chiefs of Staff in developing analytical and computer models for evaluation of strategic nuclear and conventional forces. Also the then-greatly increased Defense Department emphasis on operational test and evaluation called for IDA to develop many innovations in operational test methodology and instrumentation for air combat and air-ground combat in such projects as AIMVAL-ACEVAL. Other major areas of emphasis were assisting in coordinating, evaluating, and guiding technology-based programs, such as those in long-wave infrared sensors, materials, and propulsion.

Flax was elected to the National Academy of Engineering in 1967. He served as the Academy's home secretary from 1984 until 1992. In that capacity he was responsible for the conduct of the election of Academy members and officers and members of its Council. He was simultaneously a member of the Governing Board of the National Research Council.

Flax is a member of the Air Force Scientific Advisory Board, the Defense Intelligence Agency Scientific Advisory Committee, and a consultant to the Defense Science Board. He has been a U.S. national delegate to the NATO Advisory Group on Aerospace Research and Development, of which he is currently honorary vice chairman. He has also served on advisory bodies on engineering programs at Princeton and Stanford universities. He has delivered the Wright Brothers Lecture of the

American Institute of Aeronautics and Astronautics and the Wilbur and Orville Wright Memorial Lecture of the British Royal Aeronautical Society.

Flax is a recipient of the Lawrence Sperry Award of the Institute of Aeronautical Sciences, the Air Force Exceptional Civilian Service Medal, the Defense Intelligence Agency Exceptional Civilian Service Medal, the Department of Defense Distinguished Public Service Award, the NASA Distinguished Service Medal, and the Von Karman Medal of the NATO Advisory Group for Aerospace Research and Development. He was also a recipient of the General Thomas D. White Air Force Space Trophy. In 1992 Flax was designated an Elder Statesman of Aviation by the National Aeronautic Association.

Defense Aerospace and the New World Order

WILLIAM J. PERRY

Fundamentally the "new world order" results from the new freedom in Eastern Europe and the Soviet Union. The spirit of this new freedom is captured by a memorable statement that Victor Hugo made about 100 years ago, "More powerful than the tread of mighty armies is an idea whose time has come."

In 1989 the idea of freedom came to Eastern Europe. In 1991 it came to the Soviet Union, and in both cases it proved to be more powerful than the mighty Red Army. In each case, we have a powerful symbol of freedom. In December 1989, there was a student sitting on top of the Berlin Wall with a pickax in one hand and a bottle of champagne in the other. In August 1991, it was Boris Yeltsin standing on a tank with a microphone in his hand denouncing the coup leaders.

But this new order is more than symbols; it also has substance. Just two months after Boris Yeltsin stood on the tank, President Bush announced a unilateral reduction in the nuclear arms of the United States, a move that essentially eliminates all the ground-based and sea-based tactical nuclear weapons in the U.S. forces. A week later, President Gorbachev made a reciprocal announcement. Subsequently, we have begun discussions with the new Russian government on how to cooperate in operating ballistic missile warning systems and how to cooperate in the development of ballistic missile defense systems. How things have changed!

For more than four decades we have lived with the threat of nuclear war hanging over our heads like a dark cloud, and now that cloud is drifting away. We are truly entering a new era. This era is characterized in Russia by their second great revolution this century. It is characterized by the end of the Cold War—and confrontation with the Soviet Union being replaced by cooperative security with Russia. Paraphrasing Albert Einstein, we may now say about our national security that "everything has changed except the way we think."

How must our thinking change to accommodate the changes in the world? In this new world order, we breathe easier because of the removal of the nuclear threat, and we welcome a decline in the defense burden and the chance for a "peace dividend." But it is critical for our future how this decrease in defense spending is made. As we scale down defense spending, we need to do it in such a way that we don't cripple our ability to deal with regional military threats, that we don't lose our ability to reconstitute our military forces if a superpower threat reemerges, and that we don't damage our commercial aerospace business.

Let us consider what kind of challenges these three criteria pose. About three years ago, in an article in *Foreign Affairs*, I pointed out that the Soviet Union had a new leader who seemed to be serious about introducing new thinking to foreign policy. I suggested that this new thinking would lead to an end of the Cold War, but that it would very likely lead to an increase in regional wars; that as a result, there would be a decline in our defense budget and a restructuring of our defense forces so that they could be more effective in regional wars, and that would lead specifically to an emphasis on three technologies: stealth technology, C^3I (command, control, communications, and intelligence) technologies, and technologies associated with precision-guided weapons. The forecast at that time seemed radical. Today it seems timid. All of these things have happened. The Cold War has ended, the regional wars have started. The changes occurred sooner and more intensely than was expected when that article was written. The war with Iraq demonstrated that the central problem for U.S. security into the next century will be dealing with regional conflicts. One of the particular issues for the Defense Department is the effectiveness in regional conflicts of a military capability that was developed for an entirely different scenario, namely, war with the Soviet Union in central Europe.

PLANNING AN UNFAIR ADVANTAGE

In the late 1970s when I was in the Pentagon, and before that time, we saw the threat from the Soviet Union as a short-warning attack of armored forces in Europe, and we estimated that we would be outnumbered two or three to one in such an attack. It was politically and financially unrealistic to try to deal with that threat on a tank-for-tank basis, and therefore we developed the "offset strategy." Basically the offset strategy called for putting primary emphasis on air power for the protection of our military forces, but not simply better aircraft. The key was to give those aircraft what we thought of as an "unfair competitive advantage" arising from the support equipment that was supplied them: C^3I, defense suppression, and precision-guided weapons.

Those systems, which were developed in the 1970s and produced and deployed in the 1980s, were used for the first time in Operation Desert Storm. They demonstrated that forces with that capability have an advantage over forces that do not have it—comparable to the advantage a tank army has over a horse cavalry. In short, they simply outclass the opponents so that there is no contest.

The first critical component of this new capability is C^3I. In Desert Storm we pulled together a combination never before used in war. We used our reconnaissance satellites, which were developed originally for national intelligence, for combat support. We employed AWACS to get a continuous order-of-battle of all air vehicles, and for the first time used a system called JSTARS for a continuous order-of-battle of all ground vehicles. We made extensive use of night vision. We made extensive use of global positioning satellites to locate our forces on the ground. All of this gave our commanders superb "situation awareness"; that is, they knew precisely where enemy forces were located, where friendly forces were located, and where they themselves were located. At the same time, very early in that campaign, our forces essentially destroyed the Iraqi C^3I system. In effect, while our battlefield commanders were uniquely capable of knowing what was going on in the battlefield, the Iraqi commander's situation awareness was what he could learn by looking out of the top of his bunker.

That was the first and most crucial component of this new military capability. The second was the use of precision-guided weapons. The F-117 attack bomber conducted 1,300 sorties and dropped 2,100 laser-guided bombs. Of these, 1,700 bombs landed within 10 feet of the designated aim point and destroyed 90 percent of their targets. There is no precedent for that sort of performance in any previous use of air power.

The final component was defense suppression. To appreciate the accomplishment of our defense suppression system, it is necessary to understand that the Iraqis had a modern, dense, netted, hardened air defense system. The air defense around Baghdad in some ways was denser than the air defense around Moscow. Historically, we could expect attrition losses of 1 to 2 percent, going against that sort of an air defense. With the 3,000 sorties a day we were conducting, if we had suffered 1 to 2 percent attrition rates, we would have been losing 30 to 60 airplanes every day, and over a 30-day campaign we would have lost 1,000 to 2,000 airplanes.

Instead of losing 30 to 60 airplanes a day, we lost about one a day. We had about one-thirtieth of 1 percent attrition rate. This is a result of the introduction of Stealth (the F-117 and the Tomahawk missile) and the use of antiradiation missiles. The combination of these three essentially destroyed the Iraqi air defense electronics subsystems. As a consequence, that air defense did not have radars or command and control, and the Iraqis were left simply with guns, which they could fire visually or in a barrage. It was a combination of visual and barrage fire that gave them the one-thirtieth of 1 percent attrition results.

Equally important as these three separate components was the synergism among them. The effectiveness of the defense suppression depended on the precision-guided weapons. The effectiveness of the precision-guided weapons depended on the reconnaissance systems to know where to target. The very survivability of the reconnaissance systems depended on the defense suppression.

NEW CHALLENGES AFTER THE COLD WAR

Thus, the decisive factor in this war was an application of U.S. technological capability, especially the truly extraordinary application of air power deployed in what I would call a system of systems. This system of systems was designed, developed, and produced in the United States aerospace industry. It was not designed, developed, or produced in government arsenals. Our aerospace industry, partly because of its own success, is now in danger of being dismantled, much as we began to dismantle it in the period between the end of World War II and the beginning of the Korean War. Therefore, defense planners and aerospace industry leaders face two major challenges today. The Cold War was the stimulus for the U.S. aerospace industry's development of the world's most effective military capability; now that the Cold War is over, along with its stimulus for defense spending, will we allow our capability to deal with regional threats to erode?

The Cold War also provided the stimulus for defense R&D, which provided the technological edge for our commercial aerospace industry and the infrastructure that supported this industry and contributed substantially to U.S. world leadership in this field. Over the last decade, the commercial aerospace industry contributed more than $150 billion to a net positive balance of trade for the United States. With the defense decline, will we allow this leadership to erode?

It is fair to ask how well our present defense plan meets these two challenges facing the aerospace industry today. Simply stated, our present defense plan consists of bringing the forces down about 25 percent, maintaining R&D at nearly the present level, phasing down most of our production lines, and letting market forces determine what happens to the aerospace industry. Although this oversimplifies a complex picture, I do not think I've misrepresented it.

PROBLEMS AHEAD FOR THE AEROSPACE INDUSTRY

In my judgment, this strategy will not meet the challenges I have described. I see serious problems ahead for the aerospace industry. This strategy could very well lead to the collapse of the industrial military capability developed over the past four decades, and it could lead to some decline in our commercial capability as well.

The first problem is that the plan to sustain the R&D level will not be easy to execute. We have through the last four decades maintained leadership in high-energy physics, in manned space, in computers and software, in microelectronics and advanced materials. Defense R&D played an important role in that leadership position, but in all cases that R&D was supported and defended on the basis of the Cold War. In the absence of the Cold War and the presence of very real economic problems in this country, it will be very difficult politically to maintain defense R&D, which has supplied an infrastructure not only for the aerospace industry, but for other industries as well.

Even if we can maintain defense R&D at present levels, there is still a problem with this strategy. The problem is that R&D on the shelf is not so easy to take off the shelf. The notion that one can tie a blue ribbon around an R&D program and set it on a shelf for use when needed three or four years later is a misconception of how engineering is done. The third problem with an R&D-only strategy is that manufacturing skills are as important to our success as technology skills. I do not see any emphasis in the Defense Department's strategy on maintaining a manufacturing base.

CONSIDERING ALTERNATIVES

My consideration of alternative strategies would start by confirming the strategy of sustaining the defense technology base. It should be maintained, not because we want to put designs on the shelf, but because we want to keep engineering teams at the state of the art, and this is a way to do it. It will be difficult to justify for that reason, but that is, in my judgment, the reason to sustain defense R&D at present levels.

The reassembling and retraining of engineering teams will certainly be the longest lead time in any reconstitution of our military capability. So maintaining engineering teams at readiness is a necessary, but by no means sufficient, condition for reconstitution. In addition, it will be important to maintain a small but modern production base for systems that are "defense unique." It is easy to make up a short list of such systems: fighter aircraft, submarines, air-to-air missiles, antitank missiles, for example. Many of the components in those systems can be made in the commercial field, but the systems themselves have no commercial counterpart; therefore we cannot count on market forces to maintain them. If we do not maintain production lines at some level, we will simply forget how to build fighter aircraft and submarines. It is hard to imagine how we are ever going to reconstitute that capability in any reasonable time.

My third and final recommendation is for the government to take vigorous action to integrate the defense industrial base with the commercial industrial base. We can no longer afford the luxury of maintaining two separate bases. They have to be brought together. To do that, we have to break down three critical barriers. The first is the security barrier, the second is the specification barrier, and the third is the procurement barrier. These barriers have been set up through the decades based on the assumption that it was a good idea to have two separate bases. Now the barriers are so successful that it is almost impossible to work across them in an integrated facility. The most crucial of these, I think, is getting rid of military specifications. We have to create a single set of military and industrial specifications if there is ever any chance of reconstituting our military capability, because the base from which we reconstitute it will be our commercial industrial base, not our defense industrial base.

DIRECTIONS FOR DEFENSE AEROSPACE

What should defense aerospace companies be doing at this time? We have about twice the capacity we now need or are likely to need at any time in the foreseeable future. Therefore, there is going to be a substantial consolidation in the aerospace industry. It is exceedingly unlikely that the government is going to take any action to guide or facilitate that consolidation. Some aerospace companies will go out of business, some will merge out, some will hunker down, and some will diversify. Only the strongest will be able to diversify and continue to grow by entering other markets. It is very difficult to diversify into commercial markets because of the unique culture and the unique facilities that have been set up in defense companies. It is difficult but not impossible. There are a few rules that can guide that diversification, based on a history of previous diversification attempts. The trick is to try to generalize from the many failures and the few successes of these earlier efforts.

The first generalization is that each company needs to stick to its knitting in doing what it is really good at doing. If it is a systems company, it had better stay in the systems business. If it is a component company, it had better stay in the component business. Companies that try both to diversify into new markets and to break away from the things that they have been doing well for decades are asking for trouble.

The second generalization is that defense companies would do well to stay away from consumer product markets. Instead, they should seek markets where the customers are measured in the dozens, not in thousands or millions. That still is quite a large market. The whole infrastructure of the United States—transportation systems, telecommunications systems, air traffic control systems, energy systems, environmental systems—is made up of large projects dealing with relatively sophisticated customers.

The third generalization is that defense companies, as they go into commercial diversified markets, would do well to enter into partnerships or joint ventures with companies that already have the required marketing skills, instead of trying to do it independently.

Graham Greene once said, "There always comes a moment in time when a door opens and lets the future in." A door has opened and the future is coming in, whether or not the aerospace industry is ready for it. Our task is to get out in front to shape this future instead of being shaped by it.

The Future of Manned Spaceflight

AARON COHEN

U.S. manned spaceflight, a program whose technical heritage encompasses years of research in both aeronautics and rocketry, in February 1992 marked the thirtieth anniversary of America's first orbital mission. This milestone offers an appropriate vantage point from which to review the lessons of the past and to offer an assessment of how the future might unfold for this very visible, public enterprise of research and exploration on the space frontier.

After more than three decades of space exploration, America's spacefaring enterprise has evolved steadily to new levels of maturity and expertise in manned and unmanned operations on the high frontier. Our exploration of space has involved sending manned expeditions to study the moon, unmanned probes to complete what has been called the first preliminary reconnaissance of the Solar System, development of the world's first reusable manned orbiters, and the deployment of observatory-class spacecraft to study the mysteries of the universe. These vessels of exploration, both manned and unmanned, have improved significantly over the past three decades, benefiting from both the steady advance of new technologies and our growing maturity as a spacefaring nation.

Today that same spacefaring enterprise is undergoing subtle but fundamental changes as it prepares for more advanced and comprehensive forays on what John F. Kennedy called "this new ocean of space." In its program to sail that new ocean, the manned spaceflight community, in particular, is moving toward a space-based infrastructure in which a permanent presence in Earth orbit, habitation of the moon, and exploration of Mars will be possible. Making use of the unique environment of space is a hallmark of this planning effort, comprising a host of new technology applications to take advantage of the microgravity and ultravacuum conditions of space, as well as the abundant resources that await us on the moon. No longer limited to singular forays into space, the program will involve a wide range of missions and capabilities, all being brought to bear on the larger goal of expanding humanity's presence out into the Solar System.

Predicting the specifics of how this enterprise will evolve over the next 10 to 20 years is difficult. A similar cautionary note occurred to Neil Armstrong when he addressed a joint session of Congress a few weeks after Apollo 11. "Science has not yet mastered prophecy," he said. "We predict too much for the next year and yet far too little for the next ten." So while explicit predictions are difficult, it is possible to focus with some precision on the trends and conditions that are likely to affect the course of manned spaceflight in the next two pivotal decades.

It seems likely that Al Flax, whose contributions to aerospace the National Academy of Engineering is honoring in 1992, could identify with the notion that there are many, many variables to consider in attempting to forecast aerospace trends. That steady hum always audible during Al's 10 years with Air Force Research and Development in the 1960s was the sound of one technology after another whirring past in a continuum of new capabilities, and Al's job all that time was to sort among a vast set of options and match capabilities with requirements. Among the aerospace trends that Al Flax has witnessed in his career is the ever-advancing stream of choices and options facing program managers, both in the Department of Defense and in the National Aeronautics and Space Administration (NASA), with regard to the technological possibilities for meeting a given mission requirement.

"Consider the changes which have taken place in the past 25 years," he told the Philadelphia section of the American Institute of Aeronautics and Astronautics in 1964. "Before and during most of World War II, the only aerospace vehicle of any interest or utility to the Air Force was the subsonic airplane with a reciprocating engine power plant. Today we have in the active inventory of strategic weapons subsonic and supersonic turbojet and fanjet bombers, supersonic cruise missiles, and liquid-fueled and solid-fueled intercontinental ballistic missiles. The future holds an even greater variety of possible strategic delivery systems."

Al summarized what this meant back then, and what it means today, in an axiom that still holds true: "The management of this requires that we balance many conflicting factors and throughout there is the necessity for choice."

What Al saw in 1964 is certainly still one of the overwhelming realities of the aerospace business a quarter century later, and promises to be even more of a challenge in the future. That is the remarkable characteristic of our industry—to be swept along in a constant stream of technological innovation. Almost anything we engineers can conceive of doing is, in the simplest sense, technologically possible. Whether a project is politically feasible or economically viable or even desirable is, of course, another matter.

Al Flax's career straddled the decades of this century when the capabilities of our technology snowballed and our knowledge base increased geometrically. One of the reasons we pause to recognize his contribution today is that he was able to see the changes in our technological outlook as they were happening. During that same speech in 1964, Al said, "In the past, projections into the future have often been characterized by axioms of impossibility in the form of limitations on the future potential of science and technology that were thought to be inescapable consequences of physical laws."

By way of example, he cited the general belief during the 1930s that there was a limiting subsonic speed for aircraft, and he described how this was based on what was then known about transonic drag rise and the limitation of the powerplants of the day. Al also drew from the well of history the dubious proclamation of Lord Rutherford, one of the founding fathers of nuclear physics, who, only seven years before the first fission reaction, predicted that humankind would never harness the energy of the atom. Since that time, however—and Al was tapping into this as early as the mid-1960s—"the tendency has been in the opposite direction."

"We now tend to believe," he said, "that science and technology can (still operating within the laws of physics) achieve almost any desired result, given enough money and competent people. And it is true that the range of technological possibilities open to us is now very large and steadily growing larger. The test that we now have to apply in making decisions to initiate new developments is increasingly not 'can it be done' but 'do we want it more than something else which is also doable?'"

There, in one tidy nutshell in the spring of 1964, in the Pick-Carter Hotel in Cleveland, Al Flax put his finger on the reality of U.S. manned spaceflight—past, present, and future. There is, on the one hand, the purely technical consideration consisting of that which you can do and that which you cannot yet accomplish. On the other hand, there is the political process, which in the case of staging a program for human exploration of space, consists of that which you have the necessary national resolve to undertake and the appropriate political coalitions to accomplish.

Those forces do not tend to interact gracefully, and that is one reason why America's space enterprise is often the subject of a contentious and rather high-pitched debate. Forging a consensus on space exploration is difficult. There are many interests and priorities competing for limited resources. This country's total investment in the civilian space program since the creation of NASA in 1958 has been, in 1990 dollars, $410 billion. That is a considerable sum to be sure, and I know history will judge that it has been money well spent. The White House Office of Science and Technology Policy recently concluded that at least one-third of America's economic growth in the past 50 years has been the direct result of investment in science and technology programs. There is a payoff to the programs NASA pursues. But the payoff is not always immediate, nor is it always easy to assess. In these tight budget times, space exploration becomes an easy target for reductions.

So if we are going to address the future of manned spaceflight, we have to recognize at the outset that here is an enterprise that not only has to navigate in outer space but also has to ply that somewhat murky realm where capabilities, budgets, policies, politics, and the media all meet. If I might suggest an axiom here, it would be that the greatest challenges of manned spaceflight, now and in the foreseeable future, do not begin at the launch pad. One must get to the launch pad first.

Another truth that should be addressed in this discussion is the reality of what we are trying to accomplish with the program for human exploration of space. It is, at its heart, an engineering endeavor. Science is a part of it; scientific knowledge will come as a result of it; but first and foremost, this is an effort to engineer our way into the future. We are creating a transportation system and an exploration capability, and that is primarily an engineering task. We have a great deal more to learn and a great deal more work to do before the essential engineering challenges of spaceflight are comfortably behind us.

Assessing the future of manned spaceflight in many ways depends upon how successfully we make the case for space exploration and how effectively we argue for the necessary tools to get the job done. That is a tall order, given the nature of the debate over spaceflight.

Media coverage of the Space Shuttle Program, for instance, tends to concentrate on whether a mission departs five seconds or five days late, whether there are any aggravating technical glitches, and whether or not the flight crew is feeling well. All of these are important considerations, of course, and they are of understandable interest, but in a larger sense they are superficial to the broader panorama of what we are learning and accomplishing during these missions. There is a general lack of appreciation for the technical accomplishment involved in lifting four-and-a-half million pounds of machine straight up from a launch pad—and doing it safely—which in the minds of some are two mutually exclusive propositions.

In assessing where we are going with manned spaceflight, it is important to understand where we have been and what we have learned. Like any largescale engineering program of long duration, the manned spaceflight program is constantly building upon its own foundations. That is probably the single most compelling reason why, from a technical standpoint, we need a space station. But before considering the construction of a destination, a place to go in low Earth orbit, let us first consider how we get there.

The shuttle, which is today the planet's most reliable launch vehicle, represents essentially the third generation of U.S. manned spacecraft, but as a reusable, winged space plane, it is also a first-generation design in its own right and something of a radical departure from the designs that preceded it into space. As a flying vehicle, the shuttle is much more than simply the product of 30 years of space-age engineering. It is primarily a result of America's aviation heritage as well. In that sense, the shuttle represents a technological leap beyond the Mercury spacecraft in the same way that a Boeing 767 represents a quantum leap beyond the DC-3. It took 50 years to make that kind of leap in large transport aircraft and about 25 years to make that same kind of improvement in our manned spacecraft.

In that time, we increased the weight of the vehicle more than 75 times, expanded the habitable volume from 36 cubic feet to 1,765 cubic feet, and extended the number of engineering measurements—our insight, if you will, into how the vehicle performs—by a factor of 100. The fact that the U.S. manned spaceflight program had the tools and the knowledge to accomplish this in so short a time is the product of more than half a century of steady, prudent investment in basic aerospace research and development.

Since the first flight in April 1981, the shuttle fleet has logged more than 100 million statute miles and more than 300 days in space. The distance traveled in the shuttle program—in excess of one astronomical unit—is more than was logged by all previous U.S. spacecraft combined. Yet for all of that, we are still just at the beginning of developing a robust spacefaring capability.

It is interesting to note, as a gauge of our actual experience in space, that a total of 11,840 days have elapsed from October 1, 1958, the day NASA opened for business, until today. In all of that time, including the most recent space shuttle flight, there have been 672 days when at least one American was aloft in space. That is not an overwhelmingly impressive number when one considers the scope of the challenge ahead and the nature of the lessons that still must be learned. In all the time that has elapsed in the 33 years since NASA began exploring space, we have actually had people in space only a little more than five-and-a-half percent of the time.

Yet a fundamental truism of space exploration is that we learn by doing. Time aloft helps us hone our skills. Many of the tasks we accomplish routinely today would have been well beyond the scope of our capabilities in the 1960s. A more enduring presence in space, then, is considered to be the key to advancing many of the technologies, skills, and processes necessary to fulfill the agenda that now seems likely for manned spaceflight.

That agenda was set by President Bush in the summer of 1989 as we paused to celebrate the twentieth anniversary of the first lunar landing. In marking that occasion, the President proposed a long-range space exploration program that set three goals for the U.S. space program. The first goal is the construction of Space Station *Freedom*, to be complete by the end of this decade. The second goal is to return to the moon and establish a lunar base. The third goal is to embark eventually on manned expeditions to Mars. In that speech, what the President proposed was expanding the human presence off-planet. That is certainly something to think about.

NASA has made substantial progress in pioneering new technologies to take advantage of an expanded American presence on the space frontier. In fact there are three key technologies emerging from our research into Space Station and lunar base operations. Each of the three technology applications has the potential in the years ahead to exert a profound effect on our society and on American competitiveness in the global marketplace.

One application is in the field of medicine and involves recent advances with a device called the rotating wall vessel, also known as the "bioreactor." Originally developed to take advantage of the microgravity environment aboard the Space Station, the device has shown tremendous promise on the Earth in helping to understand and emulate tissue growth. Today, as we seek to learn about and treat diseases and disorders in the body, we are limited to a two-dimensional understanding of how cells grow and replicate in the body. We are limited by the way those cells can be grown in a petri dish, for example, or by the laborious process involved in taking skin samples and growing new skin for burn victims. With the bioreactor, we are now beginning to learn how to grow cells as the body grows them, in three dimensions and in the form of tissue, rather than simply a collection of cells. What this suggests, for the not-so-distant future, is astonishing. For example, it may even be possible to grow whole new organs, such as kidneys, for patients who need them.

These tantalizing possibilities are more than dreams. They are real. We are making great strides in the bioreactor program even as we speak, and have already conducted the first test flight in space aboard the shuttle. It is also worth noting here that the President's Council on Competitiveness last year predicted that biotechnology will grow from a $2-billion-per-year industry to a $50-billion industry within a decade.

A second promising technology would also take advantage of microgravity and the near-perfect vacuum of space to produce high-quality, thin films with immense potential to the nation's semiconductor industry and other applications. The Space Vacuum Epitaxy Center, working with NASA, is now preparing for the first flight of its Wake Shield Facility aboard the space shuttle next year. The goals of the program are to produce new electronic, magnetic, and superconducting thin films both in space and in terrestrial laboratories. The key to this approach is that it would be possible to conduct these processes in a vacuum chamber without walls. Researchers believe the approach has the potential to achieve defect-free performance in superconductor production by allowing atoms to do what atoms want to do. Defect-free chip performance implies smaller chips, and the smaller the microchip, the faster its performance. It is difficult to set limits for how this technology might be applied in the future.

This is the first program to characterize and use the ultravacuum of low Earth orbit. It is the first program to yield advanced technologies in thin-film processing. It is the first program to develop a U.S. free flyer platform for ultravacuum and microgravity applications. All the while, the worldwide semiconductor market is growing. It now stands in excess of $40 billion a year and is expected to increase by a billion dollars a year into the foreseeable future.

The third promising technology application awaits us on the moon. We know from the Apollo lunar surface samples that metals such as iron, aluminum, magnesium, and titanium are on the moon in abundance. We also know that we can process silicon and oxygen from the lunar regolith. But one other item found on the surface of the moon could ultimately surpass all of these in importance. Over millions of years, the solar wind has deposited vast quantities of the isotope helium-3 onto the surface of the moon. This isotope, which is quite rare here on Earth, has in laboratory tests shown great potential for future fusion reactions with deuterium. The energy produced from such a reaction has been comparable to those produced by reactions using tritium, but the by-products using helium-3 are completely harmless. There would be no radioactive by-product and no long-term storage and disposal problems.

Fusion, of course, is a long way off. The best guess is that we are 40 years away from a workable, widespread utility network based on fusion technology. But the fusion reaction last fall in England by a European team tells us that we are making progress. Since the best estimates are that the equivalent of one shuttle cargo bay of helium-3, using fusion technology, could provide the entire U.S. energy requirement for one year, it quickly becomes a topic for serious thought and planning.

If we are going to take advantage of these and a host of other technological opportunities on the high frontier, we need to move ahead with developing and settling that frontier. In considering the future of manned spaceflight, I would argue that 50 years from now, in the year 2042, when our lunar laboratories have grown into small settlements and when humans are beginning to fan out onto the plains of Mars, history will regard the decision by President Bush to engage the nation in a program of large-scale space exploration as having a resonance across the decades, and perhaps as being fundamentally more important than any other public policy decision of our time.

Yet the implications of all of this, technically speaking, are enormous. Shortly after the President announced these goals, I was asked by NASA Administrator Richard Truly to lead a 90-day study examining our capabilities and our options for carrying out this program. The study determined that in order to return to the moon to set up permanent scientific facilities, we will have to be able routinely to place 500,000 pounds of equipment and provisions into low Earth orbit. We will have to be able to boost that up out of Earth's gravity well and then propel it to the moon. This is the kind of task space engineers often refer to as "nontrivial." Thus far in the space age, history's two largest boosters, the U.S. Saturn V and the Soviet Energia rocket, have achieved a rated capacity of less than 250,000 pounds to low Earth orbit.

A second sobering number is apparent in scenarios for getting to Mars. For each expedition, it is estimated there will be the need to place about two million pounds of equipment and provisions in low Earth orbit before we can even start the journey to Mars. Thirty years ago, Wernher von Braun envisioned a super booster called Nova, which would have dwarfed the mighty Saturn V, and would have been able to loft half a million pounds to low Earth orbit, one quarter of what we need to get just one team of explorers to Mars.

Many of the other critical variables we face today in preparing for such endeavors grow out of such requirements and out of the options we will choose from to meet those requirements. Just as Al Flax envisioned, there are always choices to be made, and each choice will have its own ripple effect downstream in the twenty-first century. Consider some of the choices we face. In sizing the launch vehicles we are about to build, we must first decide between in-space assembly or direct transport of habitats to the surface of the moon and Mars. We have to choose between chemical, electric, nuclear, or other forms of propulsion. We have to a decide whether these missions will evolve over time or take the form of full-blown expeditions. Do we have the capability to build closed life-support systems?

How are we going to address the issue of radiation protection and shielding, and what level of weight penalties will that entail? Should we try to perfect artificial gravity for our Mars vehicles, or invest in physiological countermeasures so that humans can thrive in zero-g for two or three years?

I believe that in time the scenario suggested by President

Bush will come to pass. The economic implications alone of making use of a perfect vacuum, of mixing substances at a molecular level in microgravity, of achieving a better understanding of the human body, and of the promise of using the abundant resources out there in space will cause us, as a nation, to embark on a series of greater, more complicated efforts in space exploration.

I also believe that we as stewards of the nation's space exploration program can have some positive influence on the way in which that future unfolds. I believe the realities of the present day suggest five things we in the space program can do to help bring about the capabilities that are now on the drawing boards of NASA.

Our first task must be to tell the story of spaceflight better. We have to help the taxpayers understand what it is they are investing in, and why. We have to share our vision for the future.

The second thing we have to do is to help promote a consensus and a constituency for spaceflight and for exploration.

A third goal must be to move ahead with Space Station *Freedom* in this decade. We have to cut through the politics and the indecision and get on with it. An orbiting research laboratory is essential to our future in space. Right now, we are limited in what we can do out there on the space frontier by two things: the human body and the space environment itself. Spaceflight has a pronounced effect on the body. It affects space travelers at the cellular level, and we see changes in the heart, the lungs, the kidneys, the blood vessels, the hormone-secreting glands, and the bones. Muscles lose protein, bones lose calcium, and metabolic processes such as the production of hormones, red blood cells, and white blood cells may be altered. We cannot prudently commit to the challenges ahead until we better understand how to keep people healthy and productive out there.

We also need to learn more about space basics. Even after 30 years, we still have fundamental questions about such prosaic things as paints, coatings, metals, wiring, spacecraft design, and a whole host of other considerations. The shuttle, versatile as it is, can only do so much to help us in the learning process because its "stay time" in space is limited. For these and other reasons, whatever exploration architecture we may ultimately chose, the Space Station is the first crucial engineering step that must be taken.

The fourth goal is to reduce the cost of flying the shuttle. These are difficult financial times, and we have to seek greater efficiencies in our spaceflight operations. To that end, NASA has embarked on a five-year program to reduce the shuttle's operating costs by 15 percent. We intend to do this while also preserving our safety margins, whatever the cost. At the same time, there is a payback from flying reusable vehicles, and we must learn to take better advantage of it.

One of the main values of the shuttle is its reusability. It is that quality which has helped make the shuttle so reliable. The more we fly the orbiter fleet, the more we learn and the more understanding we have of how each system performs in a given circumstance. As we get smarter, I believe we can safely take advantage of these insights and improve the overall efficiency and operation of the system.

Finally, we must make a firm commitment within NASA to hold the line on cost growth of new programs. This is difficult to do, especially within government programs, because the uncertainties of the year-to-year congressional budget process expose new programs to a series of fits and starts. Some programs are turned off and on like a light switch for years before they finally achieve any sort of stable funding. By the same token, NASA must develop a better track record of meeting cost commitments if we are to enjoy congressional support in the future as these more costly and more ambitious projects are being debated. This is one area where we have to do better. And we will do better.

The future of manned spaceflight, while promising, faces many challenges. We see various technical barriers on the horizon, but that is the rule, rather than the exception, in the business of aerospace. In time, we will solve those problems and then move on to the next set of challenges. In 1951, when Al Flax signed his affidavit of appointment to a subcommittee of the National Advisory Committee for Aeronautics, supersonic flight was still a rarity, wind tunnels were still unable to emulate flight in that regime, and the computational capabilities of the time were, by today's standards, almost prehistoric.

It was the year that Bell Telephone gave transistors their first commercial application in a new long-distance, direct-dial telephone service; the year that Chrysler first installed power steering in 10,000 Crown Imperial sedans; the year that CBS began broadcasting in color; and the same year that a new coaxial cable carried the first transcontinental television broadcast. Today, we are accustomed to live transmissions from Neptune, delayed only by the four-hour, one-way light time to Earth.

Al Flax, and most of the rest of us here today, have been fortunate to live in a time when we see old barriers falling. We have witnessed firsthand the exciting changes that technological progress has brought. As we pause to mark the contribution Al Flax has made to aerospace, and to consider the future of this industry, we can be certain only that change will be our constant companion, and that the results of our efforts will never cease to amaze us.

Aviation: The Timeless Industry

BRIAN H. ROWE

From an engineering perspective, we may be entering the most exciting, rewarding, and interesting time during which one can work in the U.S. air transportation industry. At the same time, we may be entering a period of steadily declining U.S. leadership in civil aeronautics—if we allow ourselves to become complacent. This paradox stems from the fact that world airline economics are in dreadful shape, even though opportunities for technical advances in civil aviation abound. To maintain our leadership in this environment, we must shift our focus from "high tech" to cost-conscious designs made possible through concurrent engineering, coupled with better-than-ever customer support.

THE AVIATION INDUSTRY

For the most part, I am optimistic about the long-term future of the aviation industry, 10 to 20 years out. I believe that, despite the current problems, a more streamlined, cost-efficient industry can and will emerge out of the ashes of 1991 and 1992, yet there are still some serious financial problems that will continue at least through 1994.

In 1991 the world's airlines sustained a record loss of $4 billion. Losses by U.S. airlines accounted for half of that total. Projections for 1992 were for more of the same. An upturn was forecast for the second half, but that will only slow down the losses, not eliminate them. Commercial air traffic was down in 1991, and fuel prices were up 50 percent because of the war in the Persian Gulf. Pan Am, Eastern, and Midway ceased operation; TWA declared bankruptcy.

As a result, many aircraft orders have been canceled or postponed. Boeing cut its 737 line from 21 to 17 and then to 14 aircraft a month. Unsold aircraft stand idle in the desert. This disappointing picture is the result of global recession occurring simultaneously with the final phase of the airline shakeout that started 10 years ago. In addition, there has been a dramatic cutback in military production, which will continue and probably even accelerate over the next several years.

There appears to be no uniform national strategy to encourage and accommodate the planned, healthy growth of air transportation. For example, the United States and other large industrial nations suffer from the debilitating congestion of their airports and related infrastructure. This situation leads to excessive delays and reduces the high level of service needed to keep old customers and attract new ones. Both market development and cost efficiency are adversely affected as a result.

The United States is also showing an alarming trend toward losing its market share and cost-competitive technological edge in civil aeronautical equipment. American products have historically dominated civil aeronautics. The United States enjoys a worldwide market share of 70 percent for jet transport aircraft and 80 percent for jet propulsion. Aviation products designed and built in the United States are the bright spot for U.S. exports; the U.S. aviation industry contributed a $16-billion positive balance of trade in 1991. Yet its market share is definitely eroding. Part of the reason for erosion of American market share is that U.S. transports have lost some of the strong technical edge they once had as well as some of the financial ability to support their customers.

Airbus Industrie is a strong number-two competitor to Boeing and is growing. Their backlog in number of aircraft is double that of McDonnell Douglas. Airbus just came off their first profitable year in 20 years of business. They also have outstanding engineering, as the United States has, and we must not forget that. Their product families of A300s, A31Os, A320s, A340s, and soon the A330, are strong and competitive.

Coupled with the European success in the large aircraft market is the intense competition in the small, regional jet transport market for 50- to 100-passenger aircraft—a market dominated by foreign products such as the British Aerospace BAE 146, the Canadair Regional Jet, the SAAB 2000, and the Fokker 100.

This erosion of market share applies to propulsion as well. Rolls-Royce is building good products. Their market share may currently be less than 20 percent of the world fleet, but that is primarily because Pratt & Whitney and GE Aircraft Engines are formidable competitors with large customer bases. Pratt & Whitney and GE Aircraft Engines cannot afford to become complacent in this uncertain environment. We must continue to strengthen our ties with partners such as SNECMA, IHI, MTU, and others.

In spite of these economically difficult times and fierce international competition, air transportation is one of the world's sustained growth industries. I call it a timeless industry. In the global society we are becoming, people are increasingly relying on air transportation to cross geographical boundaries at greater and greater speeds. And aviation technology continues to evolve to meet the growing demand. During the past three decades, commercial aviation grew at an average sustained rate of 7 percent a year in revenue passenger miles, and it is forecast to continue growing at an annual rate of 5 percent for the next 20 years. Revenue passenger miles are projected to hit 2 trillion by the year 2003 and 3 trillion by 2015. Most of this increase in traffic will necessitate new aircraft, provided this new equipment is cost competitive with the old equipment currently in use. As members of the National Academy of Engineering, we must help to ensure that U.S. products maintain a significant share of this giant market. It is our game to win or lose. We have the brains; we just have to manage our resources better.

THE TECHNOLOGICAL AND MARKET CHALLENGES

I suggest three essential strategies to keep the U.S. aviation industry preeminent. First, we must lead in cost-competitive technology—and that includes product, manufacturing, and services. Second, we must continue to take the initiative in fostering international partnerships. Third, we must seek new ways to work smarter, more productively, and with greater speed. We must become more skillful with the capital and technical resources we have, because they are limited.

I firmly believe these strategies will work. The classic categories of propulsion, aerodynamics, materials, avionics, and controls have formed the foundation of the aviation industry as we know it. Technology in these areas will continue to drive our future, but not as it has in the past. During the next 30 years, we can produce some dramatic advances in subsonic transportation. But to do so, we must have cost superiority in development and manufacturing. Part of this improvement will come from smarter design, better integrated manufacturing, and improved process development. Part of it will come from advances in manufacturing technology—better use of equipment, better factories, better systems, better quality control. The computer is revolutionizing design, development, and manufacturing. Just as important, we must build on the work we have done toward creating organizations without boundaries.

The major part of future gains in development and manufacturing will come from the training and motivation of our human resources. A management revolution has already begun, and it must be accelerated. People must become an integral part of the process and the solution.

Costs are a huge driver for our customers, who must make their decisions based on bottom-line payoff. At present the U.S. aviation industries are providing good value, but we need a more compelling edge to maintain our market share and worldwide partnerships.

Substantial gains are available to subsonic aviation in environmental areas such as noise reduction and cleaner engine operation. More stringent environmental requirements are certain to be forthcoming. In addition, advanced U.S. propulsion will afford us an edge in market competition. The opportunities for improvements in propulsion are still greater than in any other technology—even greater than those for advanced aerodynamic materials and airframe structures, but we must focus on how to make the best use of our technologies to reduce total cost. Reaping these gains will help ensure continued growth of air transportation through lower fares and a better environment.

TECHNOLOGY AND MARKETS ARE INTERNATIONAL

The United States should always strive to have a competitive advantage such as we currently enjoy in subsonic technology. The Europeans and the Japanese will continue to be aggressive in this area, and we must stay ahead of them if they are to continue to be our partners rather than the reverse. With the purpose of strengthening its national position, the United States should lead the world in bringing together the world's best human and technical resources toward a goal of creating more-affordable advanced subsonic flight systems.

This strategy should be applied toward the development of the high-speed civil transport (HSCT) as well. We know that the Europeans and Japanese are moving into HSCT technology. They have strong support from their governments, and their efforts are progressing faster than ours in some key areas—takeoff lift and aerodynamics, for example.

At the same time that we are working toward increased NASA emphasis on advanced subsonic initiatives, GE and Pratt & Whitney are advocating, in the short term, passage of a NASA budget that includes HSCT program dollars to get the concept accepted. The GE and Pratt & Whitney agreement to pursue propulsion development for an HSCT is perhaps the first step in putting the national interest over wasteful, duplicative uses of resources. This will generate interest from our foreign partners that we could have a real team effort for the development of an HSCT.

A new effort of considerable magnitude and technical complexity will also attract the best and brightest people to our industry. Such an effort will help to make our industry an exciting place to work—a place in which young talent will set lofty goals as they plan their careers. As I mentioned previously, it is important to recognize that the United States is unlikely to "go it alone" in the years ahead. For the foreseeable future, international partnerships will push technology. GE's relationship with SNECMA is a good example of how teams can work together and produce cost-competitive products.

An important criterion for any partnership should be that it increases GNP, short- and long-term, for all concerned. This is the way GE and SNECMA viewed collaboration on their CFMI family of engines. The partnership has received orders for more than 10,000 CFM56 engines in the past 12 years, and we have extended the life of the DC-8 aircraft. The re-engining of the late DC-8s created jobs for GE, McDonnell Douglas, and our suppliers. In GE's case, that could include as many as 750 suppliers—smaller companies that count on us for their livelihood. Moreover, these engines gave birth to a new line of Boeing 737s and to the Airbus A320 and A340, the production of which also created jobs.

Efficient, clean engines, like the CFM56 family, in the medium-range class, helped make the hub-and-spoke strategy possible. As a result, more jobs were created. I should add that our successful military engagement—Desert Storm—was helped considerably by the vastly improved, re-engined KC-135R tankers with CFM56 engines.

When I look at international technical and economic partnerships, I see two blades of grass growing where one grew before, and that is good for everyone. Today in Cincinnati, there are Italian, Japanese, and French technologists working side-by-side with us, and we have engineers in their countries. We have teamed to protect our technology in these partnerships, and, today, SNECMA is a valued partner and not a competitor. Our partnership with SNECMA is a good example, but GE also has successful partnerships, joint ventures, and coproduction arrangements with 12 other international companies.

We have learned that we must share to gain. If we do not share, our foreign customers will go elsewhere. If we share, we will increase U.S. jobs, improve our balance of trade, and hold our market share. We should not plan on total domination of the market, however. I expect that during the next decade, the emergence of a flight system for high-speed travel will be a world system and not controlled by any one country. Such a system demands too many capital and technical resources for the United States, France, Germany, or Japan to dominate. I would like to see us lead this effort, but we must earn this right by having the technology base on which to build a superior product.

THE FLIGHT PATH TO THE FUTURE

As we look to the future, consider these two points. First, we are grossly mistaken if we start believing that profits on a sale are created in marketing. Profits are created when design, engineering, and manufacturing work together to create and execute cost-competitive technologies. The initial challenge is to make better quality products at the lowest possible cost and to deliver the best value to the customer with improved speed and quality. The resultant products must anticipate the changing dynamics of the marketplace and give a company its edge. Clearly, the customer and the marketplace dictate price today. That is true for the airlines as well as for manufacturers. If manufacturers price a flight system too high, the airlines will not buy. If airlines price the seat too high, the public will not fly. That is why the lower a company can drive down production and operating costs, the more successful it will be as a supplier.

In the final analysis, we are all suppliers. We supply technology, airframes, and engines; the airlines supply seats. The more that U.S. companies can help expand the aviation market by making travel more affordable to the ultimate customer, the more secure they will make the United States in maintaining its leadership in aviation technology.

My second point is this: Our industry needs well-reasoned U.S. economic policies. There are many ways to generate resurgence in this economy. An accelerated depreciation allowance and an investment tax credit to businesses that invest in improving productivity and quality would be good places to start. The Congress should work with the airlines and manufacturers to look for ways to make the aviation infrastructure perform better.

We need airport improvements and extensions as well as continued investment in air traffic management. We need rapid transit systems connecting airports with central cities to improve passenger access. We need to take the inconvenience out of travel. We also need investment to help improve the competitiveness of our industry.

There is no question that the years ahead can be exciting, but there are also no guarantees of the necessary economic growth. The civil transportation industry has the right products and the right talent, but industry, higher education, and government must work together to aid the growth of the economy in this sector. This requires eliminating the petty boundaries that exist.

In every one of the businesses I manage at GE—large commercial, small commercial, marine and industrial, and military engines—we are looking at a "clean sheet of paper." The new ways of operating that emerge will provide the foundation for developing the cost-competitive technology that will power us into the next century—efficient, high-quality, and reliable technology. To the engineer in me, that's exciting. To the businessman in me, that means incredible opportunities and growth ahead.

The future is opportunity, but it is also challenge. Sustained economic growth for U.S. aviation will come about only through a renewed national interest in aviation technology. As a country, we need continued investment in resources that push technology as well as a new cooperative spirit between industry, government, and universities. We need a civil aeronautics initiative that makes the aviation industry a national priority. Finally, we need the youth of the nation to be engaged by the vision and excitement that this industry holds for their future.

Higher-Order Technology: Applying Technical Excellence to New Airplane Development

ALBERTUS D. WELLIVER

There is a misconception today about what technology really is. Technology has now come to mean specific things: fault-tolerant computers, portable satellite phones, featherweight composite materials. But these individual items, however important they may be, are not examples of the true meaning of the word *technology*. Instead, these are examples of technical excellence.

Are they important? Yes. But there is a higher form of technology at work behind these specific pieces of equipment—behind the benefits they produce. The ultimate technology is not based on microchips, alloys, or composites. Instead, higher-order technology is based on human experience, wisdom, and judgment. Higher-order technology is knowing how, where, and when to apply individual instances of technical excellence. In short, the development of technical excellence is essential, but it does not mean much unless it finds its way into a product and produces a benefit.

In aviation, the ultimate technology means listening to the customer and adding value to the airplane. In Boeing's design process, every new technology development must earn its way onto an airplane by adding value in one of three ways: increased safety, improved operational efficiency and economic utility, and greater customer satisfaction.

Aviation history is full of examples of technologies that looked like winners but fell far short of expectations. Some of the examples are tragic mistakes; others are only humorous oddities, like Howard Hughes's Spruce Goose. Still the biggest aircraft ever to fly, it now sits in an amusement park in Long Beach, California. But the "Goose" was a true technical pioneer. Its compression-molded birch construction is one of the earliest examples of composite materials. Unfortunately, even with eight engines it was grossly underpowered. A single mile-long hop was all it could manage. It is a classic case of technical excellence in one area, the use of new material, but a lack of true technology in how to use it.

Higher-order technology, the technology of knowing how to apply examples of technical excellence, is now shaping the development of two exciting classes of airplane. One is the development of large subsonic transports, including the potential for new airplanes larger then any current commercial transports. The other is the prospect of a high-speed civil transport (HSCT).

Although these two categories of airplane are very different, they do have one thing in common: A successful design will depend on how technical excellence is applied to the satisfaction of the customer. By themselves, individual examples of technical excellence will not produce a winning design. A good airplane is the result of good decisions, not just good components.

LARGE SUBSONIC TRANSPORTS

When the 747 was first under development in the mid-1960s, Boeing engineers studied a number of different concepts. Some were simply stretched 707s. Others were all-new designs, including double-deck and single-deck airplanes ranging from 250 to 500 passengers in size.

The stretched 707 was quickly ruled out. It was too small to serve a rapidly expanding travel market. The decision was made to build a brand new airplane that was much bigger. The question came down to whether to build a double-deck or single-deck airplane. The choice was tough. Both had advantages and disadvantages, and both configurations featured the latest in technical excellence. This included such innovations as the first high-bypass turbofans, the most advanced high-lift flap and slat combinations, and what was then a new concept in avionics, the inertial navigation system.

But it was not any of these individual examples of technical excellence that made the 747 a success. Instead, it was how this technical excellence was applied to the airplane. If we had built a double-deck, single-aisle airplane, it would not be able to carry the 8-foot-square containers that were beginning to dominate the cargo market, and our customers told us we had to have container-carrying ability to meet varying passenger and freight demand. The double-deck layout became one wide, single deck with twin aisles. This configuration was wonderful for passengers, and it produced the finest freighter in existence.

Wisdom, judgment, and experience produced a winning design that applied the latest in technical excellence. But is was the human beings who decided how to apply it that made the difference, not the hardware itself.

Today it is hard to imagine that the 747 could ever have looked any different than it does, with its distinctive hump and graceful lines. Just picture how the airplane would look if it had been built with the original decks, single-aisles, and no hump. And just think about whether this airplane would have been successful without its wide, twin-aisle layout.

Despite the many changes in configuration, the design process for the 747 was not unique. The decisions involved in applying technical excellence are always complex. In any airplane program, many concepts must be explored before the best solution is found.

The development of the 777 is an example of how this exploration process continues to shape every design. The first 777s will enter service in 1995, becoming the largest commercial twin-jet transports. The 777 is intended to meet the market needs better than any other airplane in its class. It does this by providing new levels of passenger comfort, unsurpassed operation economics, and simplified maintenance procedures, to name just a few of the design objectives.

As with the 747, the 777 incorporates many examples of technical excellence. Included are fly-by-wire flight controls, lightweight composite materials, an advanced airplane information management system, and state-of-the-art avionics. For instance, the air data inertial reference system has 50 percent fewer parts, which improves reliability and eases maintenance.

The fundamental question, though, is how to apply these and other examples of technical excellence. When Boeing began thinking about the airplane that would become the 777, many concepts centered on modifying the 767. As surprising as it may seem now, Boeing thought about stretching the 767 and even adding a double-deck section to the rear fuselage. The airplane would have had one full-length deck and a second, half-length upper deck. Obviously, Boeing did neither of those things but instead decided that a new airplane was the best way to meet market needs.

The same decision-making process applies to all other commercial transport studies, including the potential future development of large commercial transports. Air travel is currently growing at rates that will cause passenger traffic to double by the year 2005. As a result, Boeing and others have begun preliminary studies of airplanes that could potentially be larger than the 747.

If an airplane that size is ever built, it would certainly include more than a few examples of technical excellence—things like digitally controlled high-bypass engines, extensive use of composite materials, computer-modeled aerodynamics, and advanced, integrated electronic systems. In addition, everything we have ever learned about safety, operational efficiency, performance, and customer satisfaction would be in that airplane.

As with the 747 and 777, however, success will depend not only on individual examples of technical excellence but on the higher-order process of applying it. And what will the final result be? Simply speaking, it is too early to tell. It is certain, however, what factors will shape this process. Principal among them is listening to the customer's needs and developing a keen understanding of the marketplace, including, for example, the several city pairs throughout the world where individual airlines fly back-to-back 747s within three hours of each other.

A historical look at how traffic has grown is another way to understand the market potential of an airplane bigger than the

747. The Pacific has truly become the 747 Ocean, as that airplane carries a majority of U.S.-to-Asia traffic. But on U.S.-to-Europe routes, and Europe-to-Asia routes, the evolution has been somewhat different. Although the 747 is popular, there has also been considerable growth in the use of intermediate-size airplanes.

With additional direct routes from the United States to Europe, twin-engine airplanes, especially the 757 and 767, are becoming increasingly popular. In fact, the majority of transatlantic flights are now made by twins, such as the 767. This means that any airplane larger than the 747 will have to meet Asian market needs, among others.

SUPERSONIC COMMERCIAL TRANSPORTS

The second prospect for the future is a new generation of supersonic commercial airplanes. The task here is different, but the same principles of knowing how to apply technical excellence will guide the development process.

Environmental issues are a primary focus of this process. These issues include the need to control engine emissions and the need to control noise. The top priority is gaining a better understanding of the ozone layer. Boeing simply will not build an HSCT until it is known what the airplane will do to the atmosphere. The public will not have it any other way, and neither will Boeing.

At a cruising altitude of 60,000 feet, the HSCT will fly just below the highest concentration of ozone. Oxides of nitrogen from the engines are then exhausted directly into the lower half of the ozone layer. No one knows for sure what impact this would have.

Some earlier studies indicated that a fleet of HSCTs might deplete the ozone layer by several percent, but more recent studies, based on more complex models of atmospheric chemistry, indicate that the effect on the ozone layer would most likely be much less. In fact, ozone levels could even increase.

One way or another, we have to know before an HSCT is built. That is why the research being done right now is so vital. The efforts of the National Aeronautics and Space Administration (NASA), engine manufacturers, and others will provide the information needed to design an environmentally acceptable HSCT.

The second environmental issue is noise. This includes noise from sonic booms as well as noise in a community during takeoff and landing. Currently NASA is conducting human-response studies to establish criteria for sonic boom acceptability. Without those criteria it is impossible to know what configuration will result in an environmentally acceptable airplane.

Noise is also a major factor at low altitudes. The community noise standard is clear and tough. The HSCT must be no louder than current Stage 3 subsonic transports, but no suitable engine is yet available that meets that strict standard.

With engines built before 1972, engine noise could be reduced, but a heavy penalty was paid in efficiency. Today noise can be reduced without paying as great a price in lost thrust, but the target we need to reach is still beyond today's developments. Engine noise must be reduced by 15 to 20 decibels, with less than a 5 percent loss in thrust.

A number of promising designs now being tested may allow us to reach this goal. Most of these cut engine noise by mixing low-velocity outside air with the turbine exhaust. A number of factors, however, must be balanced to apply these new engine designs successfully to a commercial HSCT.

The engine must be quiet, but it cannot be too heavy. Many promising new nozzle and suppressor concepts unfortunately impose a serious weight penalty. An effective nozzle can easily weigh as much as the power plant itself. Engine location is another critical factor. The same engine may have much different noise characteristics, depending on where it is located on the plane.

High-lift systems are a second important part of reducing ground-level noise. Vortex fences can increase lift, allowing takeoff and landing at lower body attitudes and with shorter landing gear. This allows the airplane to fly at an improved lift-to-drag ratio for climb and approach.

Wake vortex image representation in computational fluid dynamics (CFD) allows for improved evaluation of the flow characteristics that lead to better performance. This performance includes benefits such as reduced noise on the ground and less fuel consumption. This will be crucial if economic viability is to be achieved.

Many of the application judgments regarding technical developments will be made in view of the HSCT's economic constraints. Simply put, ticket prices cannot be at Concorde levels. Optimistic assessments show that if fare premiums for HSCT flights average 20 percent over the fares of subsonic flights, then nearly 65 percent of the potential HSCT market could be obtained. If fares averaged only 10 percent over, then nearly 85 percent of the market could be obtained.

The result is that for the HSCT to be economically viable, technical excellence must be achieved in two aerodynamic fields: computational fluid dynamics, or CFD; and hybrid laminar flow control (HLFC). With CFD analysis we are getting a better look at supersonic flight than we have ever had before. Computer-generated images give the clearest picture yet of supersonic flow, making possible the analysis of airflow both on the surface of the airplane as well as away from the vehicle. The result is much greater technical knowledge of shock waves produced by the wing, fuselage, and nacelles.

Hybrid laminar flow control is another area in which technical knowledge is growing rapidly. A 757 demonstrator, built as part of a Boeing/NASA project, recently became the first full-size commercial transport to fly with an HLFC system. The system is simple but effective. Air is drawn through microscopic holes in the leading edge by an engine-powered pump. The result is laminar flow over more than half the wing chord. In the spring of 1992, Boeing began working with NASA, Rockwell, and McDonnell-Douglas in testing a supersonic HLFC system. A NASA-owned F-16XL, with a cranked-arrow delta wing, will be fitted with a special HLFC glove.

If applied to an HSCT, the gains could be considerable. At Mach 2, skin friction accounts for 40 percent of drag, and an HLFC system could cut that figure substantially. At supersonic speeds, HLFC has the potential to be even more effective than at conventional speeds. Because of the greater amounts of fuel needed at Mach 2, even a small percentage reduction in drag could pay off economically.

By themselves, however, analytical tools like CFD, and new systems like HLFC, will not make the HSCT a success. It will be up to engineers, to people, to make decisions on how to apply these and other technical developments. Their choices will be based on whether value is added to the airplane. They will listen closely to the customer to ensure that market needs are met.

Higher-order technology is more than simply having the tools to get the job done. It is having the wisdom, judgment, and experience to know how to apply those tools to produce a benefit for the customer.

The Future of Rotary-Wing Aircraft

RICHARD M. CARLSON

James Burke, in his celebrated Public Broadcasting System television series, "Connections", insists that "we cannot know where we are going unless we know where we have been." In the case of rotating wings, the history dates back to the early Chinese top, Leonardo da Vinci's classic sketches, and more recently to Sir George Cayley's helicopter models. But it is generally accepted that the rotorcraft equivalent to the Wright Brothers' era started in 1923 with Juan de la Cierva, at Getafe, Spain. Cierva was attracted to the unpowered, autorotating, wing because of its lifting potential at very low forward speeds. He was motivated, for safety reasons, to provide fixed-wing airplanes with a simple device that would eliminate stall and provide the capability to land safely in the event of engine failure. Cierva called his original configuration a gyroplane, which later became known as the autogiro. The whole of Cierva's effort can be appreciated by the fact that between 1920 and 1930 he developed, and in most cases flew, some 30 different experimental autogiro types. This effort resulted in production licenses to a number of foreign countries and the production of some 480 autogiros between 1924 and 1944, 185 of which were the final model C-30.

Meanwhile the vertical takeoff and landing (VTOL) enthusiasts had been hard at work developing a powered rotor, engine, and flight-control system that would accomplish manned VTOL flight. This was no simple task since the engines and speed reduction transmissions were very heavy and the powered rotors were required to provide lift, propulsion, and control of the aircraft about all three axes. Further, the powered rotor configurations were inherently unstable. It is generally acknowledged that the first successful helicopter flight was accomplished by the co-axial Breguet-Doran on June 26, 1935. This was followed one year later by the Fock-Achgelis, side-by-side rotor, model FW-61, flown by Hanah Reich in Berlin. This helicopter activity, especially the FW-61 in Nazi Germany, created great concern in the U.S. Congress, which in 1938 appropriated $2 million for rotary-wing R&D. This national interest stimulated Igor Sikorsky and the United Aircraft Corporation to develop and fly in December 1941 the VS-300, the first helicopter with a single lifting main rotor and a single tail rotor. This configuration was then applied to a U.S. Army/Air Force design requirement, resulting in the Sikorsky Model R-4 with a larger rotor and engine. The model R-4, as shown in Figure 1, was subsequently placed in production in 1942. During World War II, some 385 R-4, R-5, and R-6 helicopters were produced for use by the U.S. Army.

At the conclusion of the war, a number of other pioneering

FIGURE 1 Sikorsky Model R-4, the first production helicopter.

FIGURE 2 Piasecki XHRP, the first tandem helicopter.

efforts were under way in the helicopter development arena, one of which was evolving at Piasecki Aircraft, now Boeing Helicopter Company. Piasecki had built and flown a small single-rotor helicopter similar to the R-4B and was in the process of developing the world's first tandem helicopter. It was at that time, in 1945, that Alexander Flax joined the Piasecki group to assist in designing and flying tandem helicopters like the XHRP shown in Figure 2. It did not take Al Flax long to discover that rotary-wing aircraft are different from fixed-wing airplanes. For one thing, 40 percent of the empty weight of a helicopter rotates and the lifting rotor in forward flight experiences in one revolution, as broad an aerodynamic, structural dynamic, and aeroelastic environment as a fixed wing sees in its entire lifetime. Al Flax absorbed all of this challenging physics and decided to document some of it. In 1947 he published the classic expository paper on "The Bending of Rotor Blades" in the *Journal of the Aeronautical Sciences.* Four years later, with coauthor Len Goland, he published a second classic paper dealing with the dynamics of rotor blades. This first paper was my introduction to the rotary-wing community, 42 years ago, when I reported to the Hiller Helicopter Company for work as a stress analyst. The chief engineer, Wayne Wiesner, handed me a copy of the paper and said "read this, Dick, and you will be the Hiller rotor blade structural expert." Twenty years later I met Al when he was at the Institute for Defense Analyses and was serving on the "blue ribbon" review board for the Lockheed AH-56 Cheyenne helicopter.

During World War II the U.S. and British governments purchased some 400 Sikorsky R-4, R-5, and R-6 helicopters, which were used for observation and reconnaissance. This experience provided the catalyst for commercial and military exploitation of the helicopter's unique VTOL and hovering capabilities. Today, logistic supply, air assault, antiarmor, medical evacuation, air-sea rescue, and antisubmarine applications in the military forces are as common as traffic report, disaster coverage, emergency medical service, offshore oil support, and executive transportation applications are in the commercial sector. Today 10 countries produce small, medium-size, and large helicopters for both commercial and military applications, and a number of other producer countries are coming up fast.

The first R-4B helicopter could hover at about 2,000 feet on a standard day in-ground-effect (IGE), carry a crew of two, and fly at 70 miles per hour for two hours with a 50-pound payload (the parachutes). Needless to say, the performance of today's helicopters has improved dramatically, which accounts for their current broad acceptance and application. To illustrate both the magnitude and the source of this improvement in performance, this paper follows a format used by George Schairer in a recent paper entitled "On the Design of Early Large Swept Wing Aircraft." He traced the evolution of the three basic elements of the Breguet Range Equation (Figure 3), drag efficiency (L/D), fuel efficiency (η_p/c), and weight efficiency (W_1/W_0), and reviewed the advances in technology that provided for the introduction of the Boeing 707 jet transport. Expanding on this format, Figure 3 portrays the mission segments that must be considered, in addition to range, to exploit VTOL capability. Vertical takeoff and landing must be available over a broad range of altitudes and ambient temperatures. The installed power (power loading, l_{p0}) required to achieve this broad range is a function of rotor aerodynamic efficiency (figure of merit, M_f), rotor disk loading (w, a design parameter) and engine lapse rate characteristics. Many VTOL missions (e.g., military, rescue, logging) require that significant time be spent in hovering flight. As with power loading, the fuel required for this flight mode is a function of rotor aerodynamic efficiency and disk loading and, as in the range segment, a function of fuel and weight efficiency. For purposes of this presentation, weight efficiency is expressed in terms of ϕ_E and

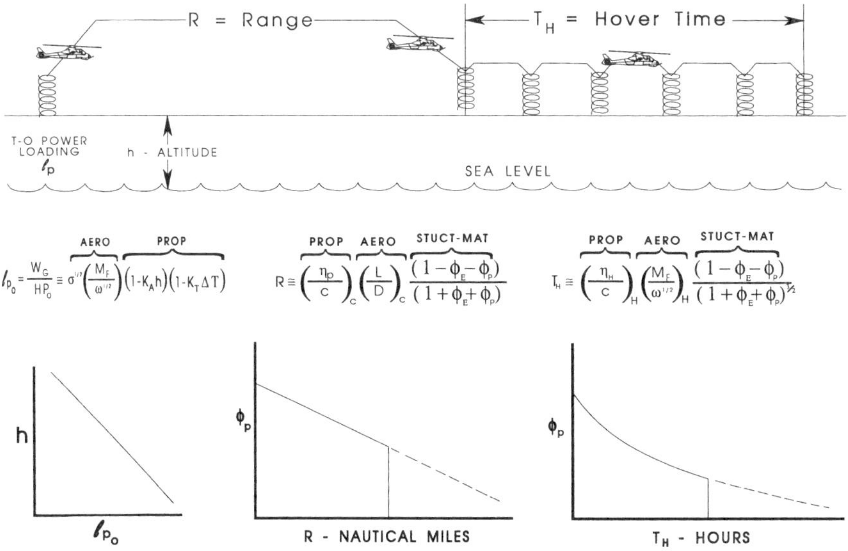

FIGURE 3 Altitude, range, and hover, mixed-mode missions.

ϕ_p, which are the empty weight (W_E) and payload (W_p) fractions of gross weight (W_G).

Of course, advances in many technologies are responsible for the enhanced status of the helicopter today. It is important to acknowledge contributions from the fields of dynamics, flight control, acoustics, safety, simulation, and above all mathematical modeling and computational methods. But, in this paper, let us address only those technologies associated with the three mission segments presented in Figure 3, which are more easily quantified in terms of vehicle performance (i.e., propulsion, aerodynamics, and materials/structures).

ADVANCES IN PROPULSION TECHNOLOGIES

Since the introduction of the R-4B, the most significant technical event during the past 50 years has been the arrival of the turboshaft engine. The impact of changing from the reciprocating engine to the gas turbine is illustrated in Figure 4 for three powerplants originally designed to deliver 800 shaft horsepower. The R-1300 powered the Sikorsky H-34, the T-53 is in the Bell UH-1 and AH-1, and the ATDE demonstration engine provided the technology for the T-800 gas turbine, which will power the new Army Comanche helicopter. The initial reduction in size and complexity provided by the gas turbine is due to the application of turbomachinery concepts to the Brayton thermodynamic cycle, resulting in much higher mass flow per unit volume in the turbine than in the reciprocating engine. Figure 5 presents historical trends in specific fuel consumption and specific weight for reciprocating and gas turbine engines used in production helicopters. The impact of the change over to gas turbines can be clearly seen. Although the specific weight dropped dramatically, the specific fuel consumption increased because materials limit the peak temperature of the Brayton thermodynamic cycle. This cycle is also more sensitive than the reciprocating engine cycles to inlet air temperature. It is interesting to note the continued improvement in specific fuel consumption and specific weight of the gas turbine. This progress is primarily due to a combination of increased efficiency of turbomachine components, increased compression ratios, and materials that allow for higher operating temperatures.

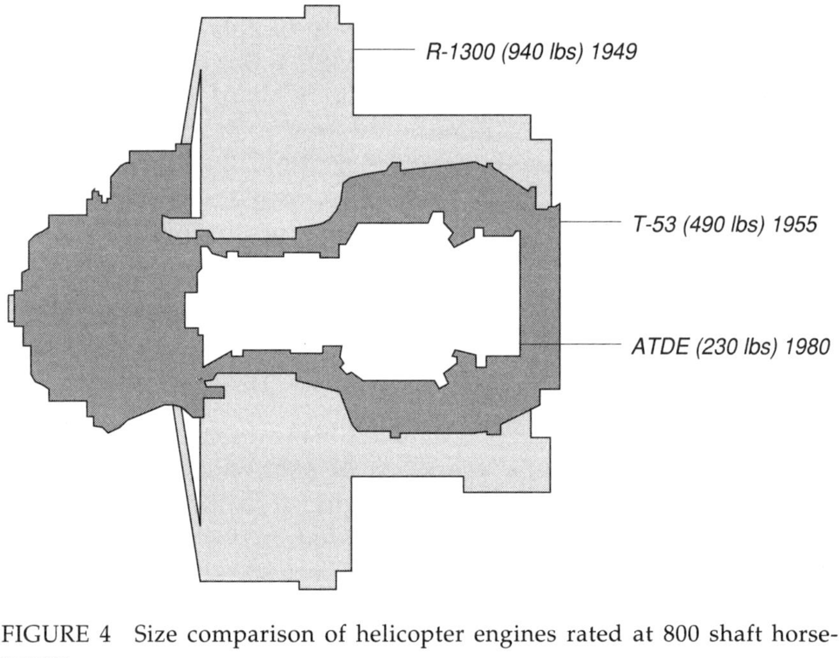

FIGURE 4 Size comparison of helicopter engines rated at 800 shaft horsepower.

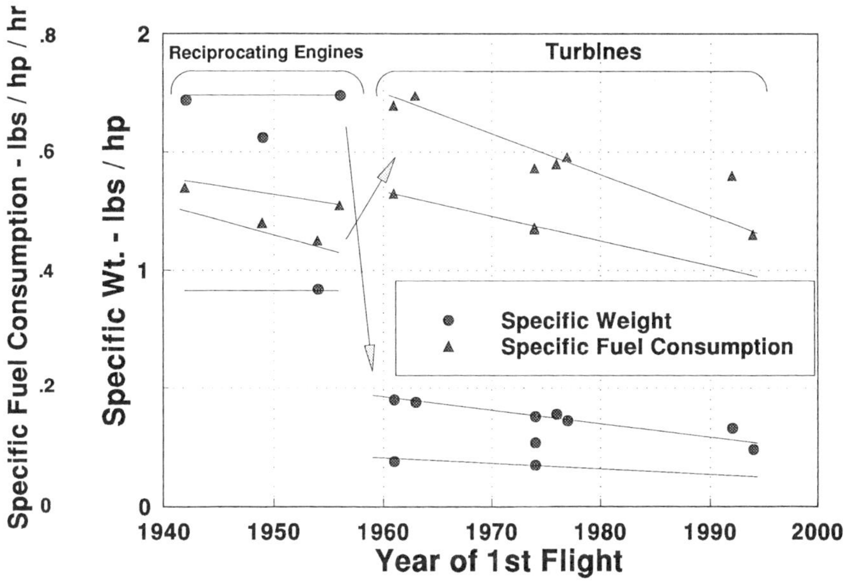

FIGURE 5 Trends in specific fuel consumption and specific weight.

IMPROVEMENT IN ROTARY-WING AERODYNAMICS

For many years, with a few exceptions, the subject of rotorcraft drag efficiency, or *L/D* ratio, has been a secondary design consideration. This has been due to the fact that forward speed performance was generally limited by the aerodynamics of the main rotor, which produces airframe vibration, and by a preponderance of low-speed mission requirements. Nevertheless, there have been improvements in rotorcraft drag efficiency, as indicated in Figure 6. These improvements have resulted from "cleaner" power plant installations and enhanced understanding of rotor-airframe interference drag.

The majority of rotorcraft aerodynamics research has been directed to the main rotor. The early rotor blades (see Figure 7) were aerodynamically constrained by manufacturing, cost, and control load considerations. The early rotor blades used symmetrical airfoil sections of constant thickness-to-chord ratio with linear twist variation along the blade radius. These blades were primarily focused on improving lift efficiency in hover flight, with forward flight performance as a "fall-out." With acceptance of hydraulic "booster" controls and some im-

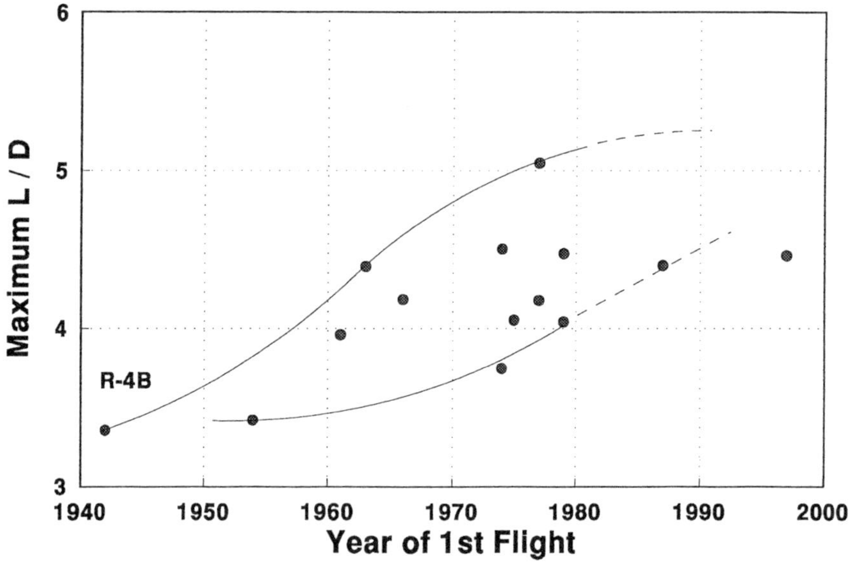

FIGURE 6 Trends in rotorcraft drag efficiency.

provement in manufacturing techniques, unsymmetrical (cambered) airfoils and variable thickness/platform tip sections were incorporated, which improved both hover efficiency and forward flight capability. A third generation of main and tail rotors started to emerge in the late 1960s with the advent of reliable fibrous composite materials, initially S-glass and later graphite and Kevlar. Application of these materials and their flexible manufacturing techniques has allowed the rotorcraft aerodynamicist to develop new airfoil sections and variations in blade geometry that improve high-speed as well as hover performance. Figure 8 presents trend data for improvements

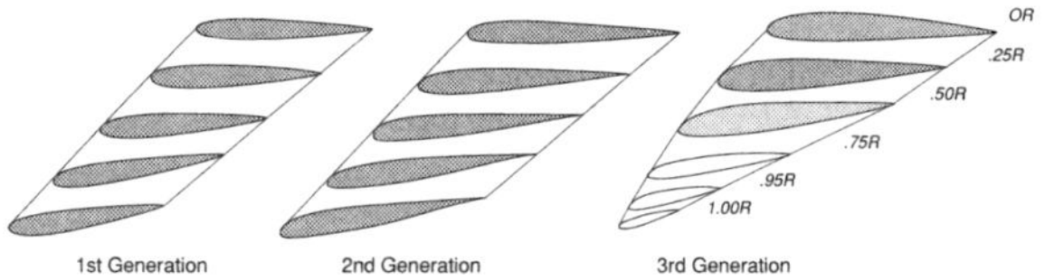

FIGURE 7 Advances in rotor geometry.

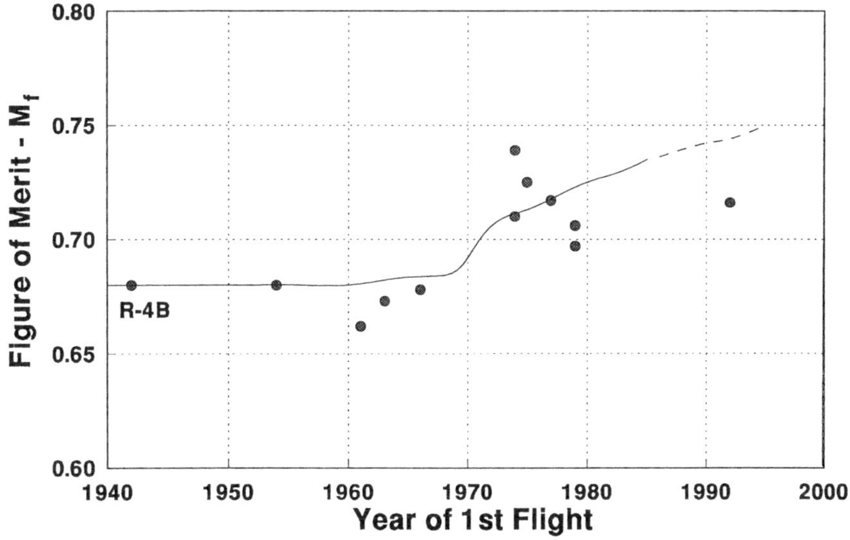

FIGURE 8 Improvement in rotor figure of merit, M_f.

in rotor figure of merit, M_f, which plays a significant role in two of three rotorcraft mixed-mode mission segments.

NEW MATERIALS AND STRUCTURES

Traditionally, we think of the application of new aerospace engineering materials in terms of saving weight, that is, reducing empty weight fraction. However, for rotorcraft and the application of fibrous composites, it was the promise of improved aerodynamic efficiency that proved to be the catalyst. In fact, application of these materials to rotor blades has not reduced rotor weight, since a rotor system mass moment of inertia requirement effectively eliminates any weight benefits. Notwithstanding the absence of rotor blade weight reduction, a second significant benefit has resulted from the use of fibrous composites in rotor blades. Whereas the replacement lives of metal components were notoriously low, the benign fatigue failure characteristics of composite materials have virtually eliminated the need for periodic replacement of rotor system components and, in addition, have enhanced the ballistic survivability of these components. Virtually all modern

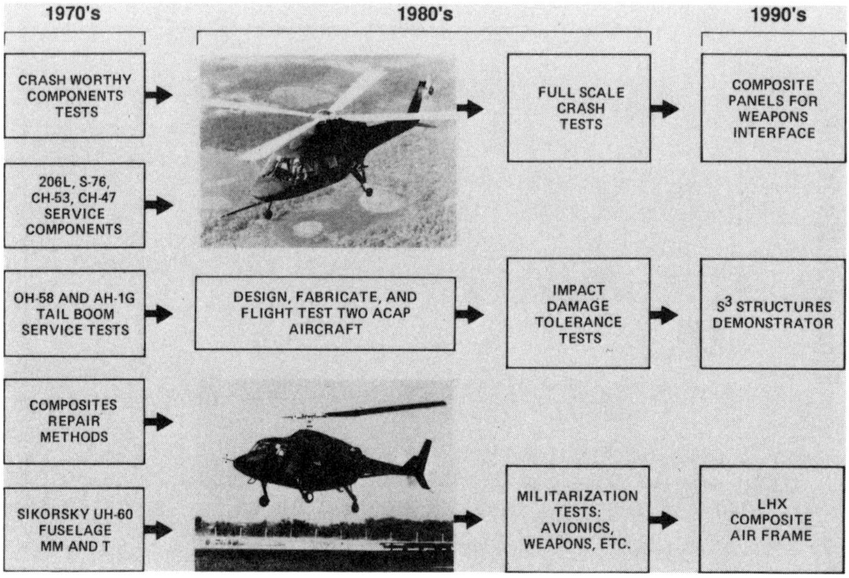

FIGURE 9 Composite materials R&D programs.

helicopters and many older models today employ fibrous composite blades.

Having been successful with rotor blades, the rotorcraft structures engineers turned their attention to the airframe, since these composite materials still offered significant potential weight and cost savings. In the 1970s (see Figure 9) the U.S. Army and NASA initiated a number of rotary-wing composite airframe component R&D projects to establish the feasibility of 10-35 percent reductions in weight and cost. These reductions were so attractive that, in 1982, the U.S. Army initiated a major composite airframe program (ACAP) to integrate the results of the component projects into a complete rotorcraft airframe to validate the weight and cost predictions. Contracts were awarded to Sikorsky Aircraft and Bell Helicopter to design, fabricate, and flight test a helicopter with a complete composite airframe. The two major goals were to demonstrate a 22 percent reduction in weight and an 18 percent reduction in cost relative to an "all metal" baseline. Both contractors met or exceeded these goals. These two programs provided the technology base and confidence level for the use of composites in the new Army RAH-66 Comanche helicopter. Figure 10 presents the rotorcraft trends in empty weight fraction, ϕ_E, and clearly shows the impact of the gas turbine. The engine weight fraction, ϕ_e, was reduced from approximately 0.12 to 0.04, or a change of 0.08, while the total empty weight fraction was reduced from approximately 0.70 to 0.55, or a change of 0.15. This reduction in empty weight fraction represents a 100 percent increase in useful load fraction for modern helicopters relative to the R-4B. It is interesting to note that, while further reductions in gas turbine specific weight (see Figure 5) have occurred since their introduction, they do not appear in the engine weight fraction trends. This is consistent with the power and disk loading trends shown in Figure 11, since decreased power loading increases empty weight and increased disk loading decreases empty weight. These trends indicate that the certificating agencies (Department of Defense and Federal Aviation Administration) and the user community have preferred to use any net weight reductions that have occurred to improve altitude and temperature performance. In any event, the empty weight trend has remained essentially level for the past 25 years (see Figure 10), and the potential empty weight fraction reductions due

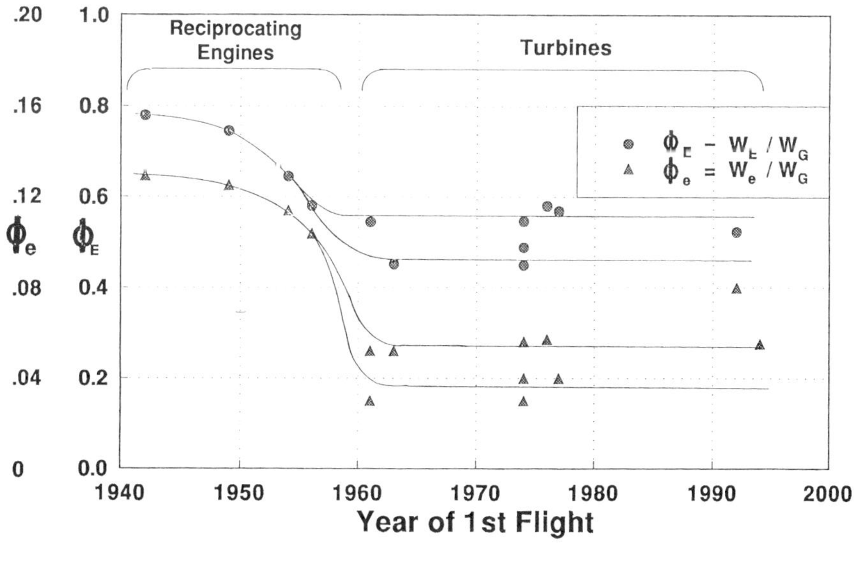

FIGURE 10 Trends in empty weight fraction, ϕ_E, and engine weight fraction, ϕ_e.

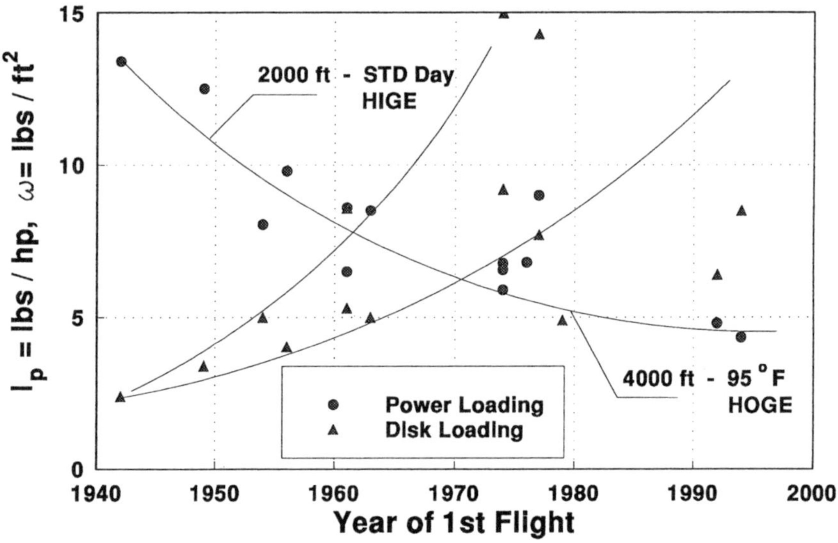

FIGURE 11 Trends in rotor craft disk loading and power loading.

the application of composites to rotorcraft airframes are yet to be realized in production vehicles.

Returning to Figure 3, we have now quantified the vehicle parameters that affect rotorcraft performance in the altitude-range-hover time segments of VTOL mixed-mode missions. Combining the trend changes in rotor efficiency (M_f), drag efficiency (L/D), fuel efficiency $(\eta/c)_H$ and $(\eta/c)_p$, and weight efficiency, ϕ_E, we may now summarize the status of today's modern rotorcraft relative to the 50-year-old R-4B. This comparison is presented in Figure 12, which indicates a substantial increase in altitude vertical takeoff performance from 2,000 to 10,000 feet and a typical range-payload design point with 3 times the range and 2.75 times the payload. The modern helicopter hover time capability is preserved only by virtue of the greatly increased payload weight fraction, the total hover time rotor efficiency of the modern helicopter being less than the R-4B because of its larger disk loading.

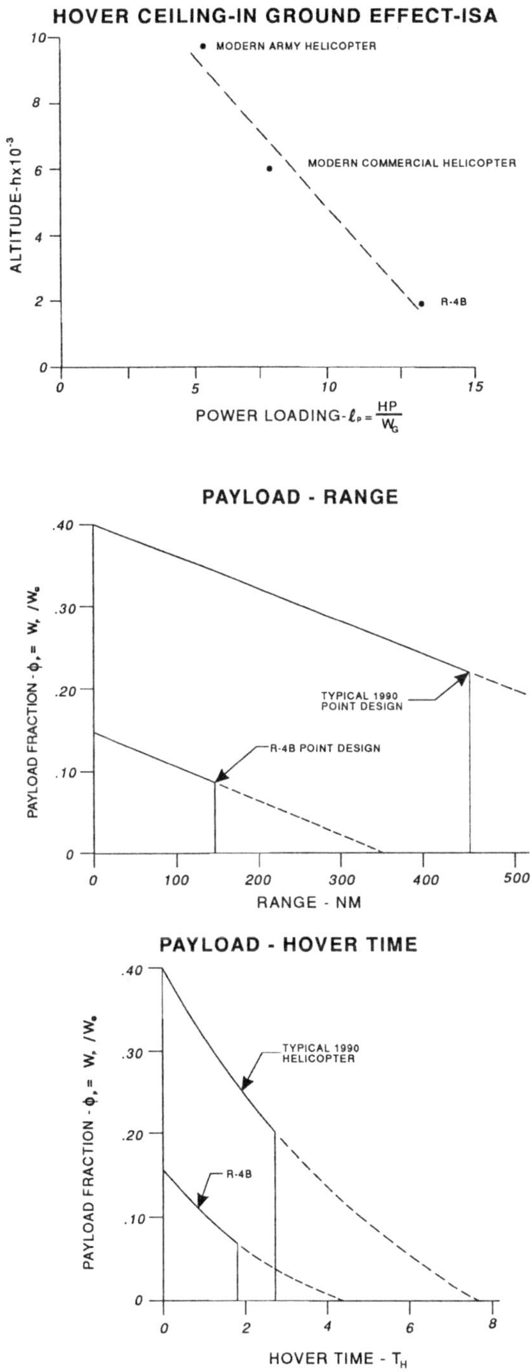

FIGURE 12 Altitude, range, and hover performance comparison.

THE ECONOMIC OUTLOOK FOR ROTORCRAFT

What does the future of rotorcraft, and its attendant technology, offer as an economic motive for those involved in the development and operation of VTOL aircraft? First, one must consider the fact that since the U.S. Army's initial purchase of 385 helicopters, its inventory grew to 8,454 vehicles in 1990, and U.S. Department of Commerce statistics indicate that since 1950 the U.S. helicopter industry has produced some 19,000 helicopters for the civilian market. The Soviet armed forces helicopter inventory stands at 4,000 today, and European production since 1950 totals approximately 15,000 vehicles, half of which have been produced by Aeroepatiale in France.

Clearly a competitive VTOL market has been established, and new helicopter projects are now emerging in the United States, France, Germany, Great Britain, Japan, China, Singapore, and India. These projects are all providing increased forward speed, as indicated by the trends in Figure 13. Increased cruise speed in turn relates to increased productivity, which is generally defined as the product of payload and speed divided by empty weight (see Figure 14).

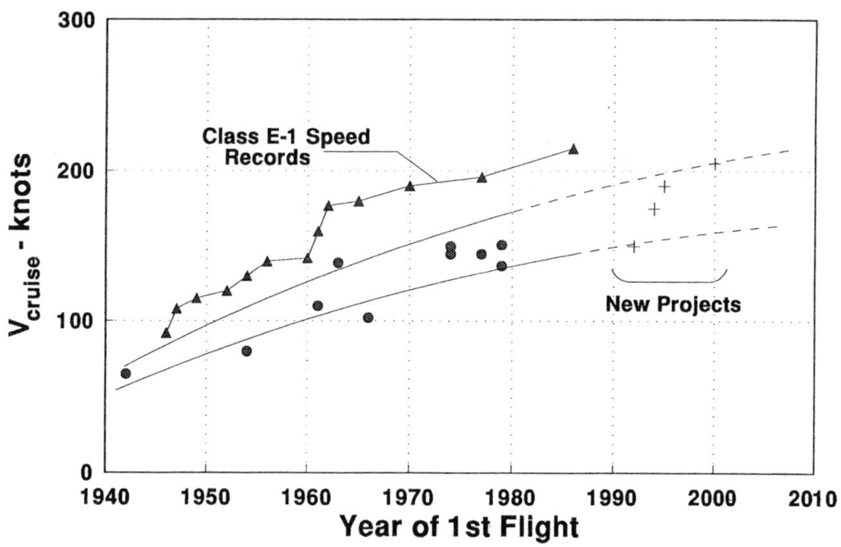

FIGURE 13 Trends in helicopter cruise speed.

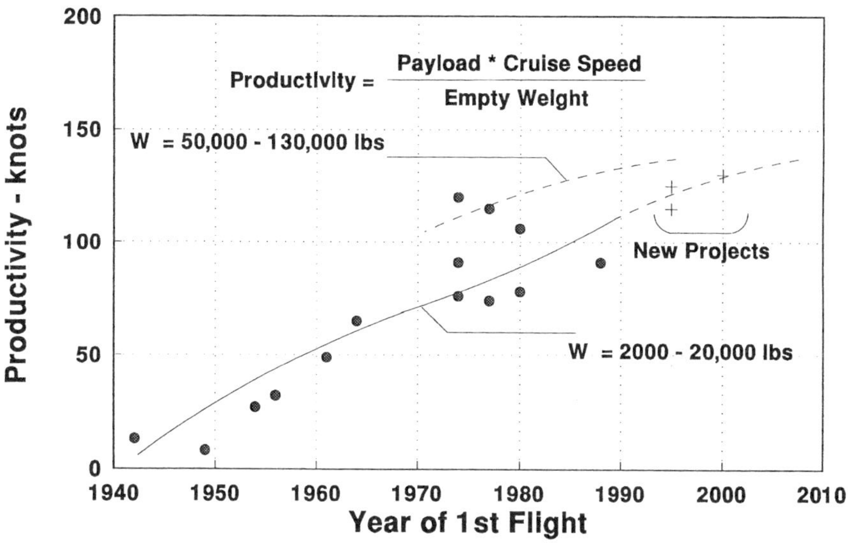

FIGURE 14 Trends in helicopter productivity.

But what, if any, are the future commercial and military markets, and what is the rotorcraft industry response to these markets? The most challenging new market emerging from the commercial sector is coming from the Pacific Rim. The Japanese government has begun a program to establish an island network of heliports as an integral part of that nation's planned transportation system. Japan refers to VTOL as the Fourth Revolution, after rail, auto, and airplane. Figure 15, reproduced from the report of a study conducted by Japan Heli Network Co. Ltd., shows the projected extent of the heliport network by the year 2020. Work has begun on the first 10 heliports, and golf courses appear to have higher priority than cities. One can extend Japan's planning process south to Indonesia, Malaysia, and Singapore, and it is clear that the entire western portion of the Pacific Rim is a latent market for high-productivity VTOL aircraft.

The outcome of recent events in Southwest Asia will most certainly support similar high-performance rotorcraft developments. The Desert Storm experience was the first real-life test of the U.S. Army Airborne logistics, assault, and antiarmor assets in a combined arms role. Figure 16 presents a dramatic

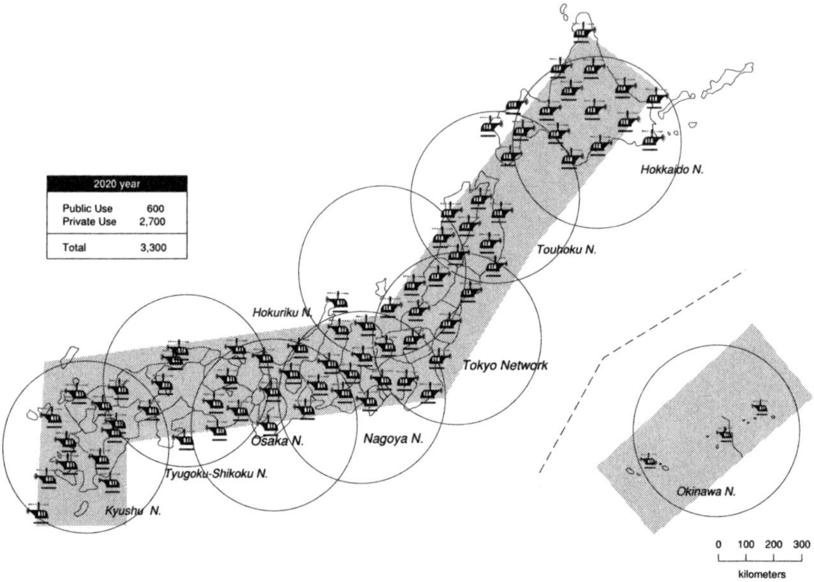

FIGURE 15 Japan's projected heliport network.

view of what organic airborne capability can produce. The heavy dashed lines indicate the positions of multinational forces 12 hours and 48 hours after combat operations began. Desert Storm has been referred to as "a high-technology test bed" and a validation of emerging army doctrine called ALO, for air-land-operations. This new doctrine envisions changing the Army Corps area of operations from a European defensive scenario with dimensions of a 50-kilometer front and a 180-kilometer depth to a more global capability with Corp dimensions of 100 kilometers by 450 kilometers. Again, it is clear that this dramatic change invites the development of new high-performance VTOL aircraft.

Since the 1950s the rotary-wing community has been actively pursuing alternate VTOL configurations that produce performance improvements beyond those of the helicopter, and these efforts are now showing signs of success. Tilt-wing performance increases were clearly demonstrated in the 1960s with the XC-142, and more recently tilt-rotor developments such as the XV-15 and the V-22 have confirmed (see Figure 17) substantial performance improvements. Recent NASA high-speed rotorcraft studies, based on past experimental flight tests and many wind tunnel tests, clearly indicate these configurations are capable of producing substantially greater productivity and range than the conventional helicopter.

New commercial projects have been initiated to match the capabilities of these advanced rotorcraft configurations to new emerging markets. Three of these projects are the European EUROFAR tilt rotor, the Japanese Ishida tilt wing, and the U.S. National Civil Tilt Rotor initiative (Bell-Boeing), which is led by the FAA Vertical Flight Program with strong support by NASA and Congress (see Figure 18).

Fifty years of research and development in the rotary-wing field have produced numerous improvements since the first flight of the R-4B. The helicopter today is no longer an oddity but an integral part of our daily life. The desire to take off and land vertically and fly in the earth's boundary layer has per-

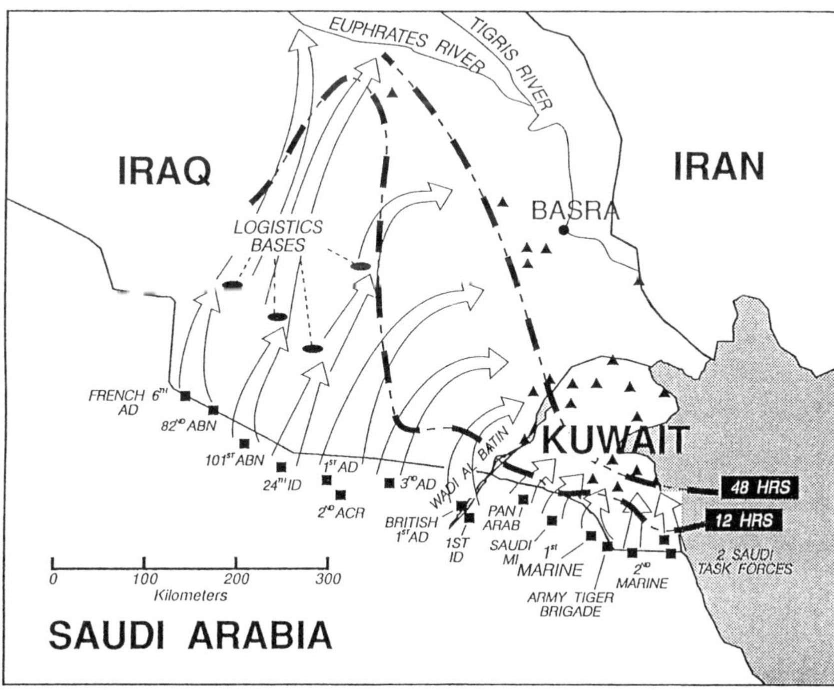

FIGURE 16 Positions of multinational forces in Operation Desert Storm.

FIGURE 17 Tilt-rotor developments XV-15 (left) and the V-22 (right).

FIGURE 18 New tilt-rotor and tilt-wing developments: (left) Eurofar; (center) Ishida; (right) Bell-Boeing.

sisted, and technology and dedication have accommodated that desire. We have seen that the advent of the turbine engines, application of composite materials, and advances in rotary-wing aerodynamics have had a profound impact on the utility and acceptance of the helicopter. The future of the helicopter is clearly assured, and advanced rotary-wing aircraft, with an additional degree of rotation, are ready to extend the performance and utility of the helicopter.

Authors

H. GUYFORD STEVER (symposium chairman), a member of the Carnegie Commission on Science, Technology, and Government, has had a career as a scientist, engineer, educator, and administrator. He holds an A.B. from Colgate University, a Ph.D. in physics from the California Institute of Technology, and numerous honorary degrees. In the past decade he has been a director of TRW Inc., Schering-Plough Corporation, and Goodyear Tire and Rubber Company; a trustee of Woods Hole Oceanographic Institute, of Science Service, and of Universities Research Association; and foreign secretary of the National Academy of Engineering. He was science and technology adviser to President Ford, director of the White House Office of Science and Technology Policy, and an ex officio member of the President's Commission on Science and Technology. He was director of the National Science Foundation and, concurrently, science adviser to Presidents Nixon and Ford. He has also been a member of the National Science Board. Before his government service, he was president of Carnegie Mellon University during the period when the Carnegie Institute of Technology and Mellon Institute were merged. He was professor of aeronautics and astronautics at Massachusetts Institute of Technology as well as head of the departments of mechanical engineering and naval architecture and marine engineering. He is a member of the National Academy of Engineering, the National Academy of Sciences, and the American Academy of Arts and Sciences; a fellow of the American Physical Society, the American Institute of Aeronautics and Astronautics, the Royal Aeronautical Society, and the Royal Society of Arts; a foreign associate of the Japan Academy of Engineering and a foreign member of Britain's Fellowship of Engineering. He has received the President's Certificate of Merit for his work in World War II, the Commander of the Order of Merit of Poland, Distinguished Public Service Medals of the Department of Defense and of the National Aeronautics and Space Administration, and other honors.

RICHARD M. CARLSON is chief of the Advanced Systems Research and Analysis Office at Ames Research Center in Moffett Field, California. He earned a B.S. degree in aeronautical engineering from the University of Washington, an M.S. degree in aeronautical engineering from the University of Seattle, and a Ph.D. degree in engineering mechanics from Stanford University. Before his career with Ames Research Laboratory, Dr. Carlson was employed in the aircraft industry for 24 years, during which time he was chief aerostructures engineer at Hiller Helicopters, Inc., and rotary-wing advanced design division engineer at the Lockheed California Company. He has been awarded three Army Meritorious Civilian Service awards and a Presidential Rank Meritorious Executive SES Award. He is a member of Sigma Xi and the Swedish Society of Aeronautics and Astronautics. He is also a fellow in the British Royal Aeronautical Society, an honorary fellow of the American Helicopter Society and a recipient of that society's Klemin Award, a fellow in the American Institute of Aeronautics and Astronautics, and a member of the National Academy of Engineering.

AARON COHEN is the director of the Lyndon B. Johnson Space Center of the National Aeronautics and Space Administration (NASA) in Houston, Texas. Cohen received a B.S. degree in mechanical engineering from Texas A&M University and an M.S. degree in applied mathematics from Stevens Institute of

Technology. He has taken advanced graduate study in mathematical physics at New York University and the University of California at Los Angeles. He received an Honorary Doctor of Engineering from Stevens Institute of Technology and an Honorary Doctor of Humane Letters from the University of Houston. Mr. Cohen's career at the Johnson Space Center began in the Apollo Spacecraft Program Office. He managed the hardware and software designed to provide guidance, navigation, and control for both the Command and Service Module (CSM) and the Lunar Module. He served as chief, Systems Integration Branch/Systems Engineering Division, and chief, Command Service Module Project Engineering Division. Mr. Cohen was also the manager for the Command and Service Modules unit of the Apollo Spacecraft Program. Prior to these assignments, he was a microwave tube design engineer at RCA and a senior research engineer at General Dynamics Corporation. Mr. Cohen is a fellow of the American Astronautics Society and a fellow of the American Institute of Aeronautics and Astronautics. His NASA awards include two Exceptional Service Medals, two Outstanding Leadership Medals, three Distinguished Service Medals, and an Engineer of the Year award. He is a member of the National Academy of Engineering.

WILLIAM J. PERRY is the deputy secretary of defense. He was formerly the chairman and chief executive officer of Technology Strategies & Alliances (formerly H&Q Technology Partners, Inc.) in Menlo Park, California, and also served part-time at Stanford University as a professor in the School of Engineering and as codirector of the Center for International Security and Arms Control. Dr. Perry received his B.S. and M.S. degrees from Stanford University and his Ph.D. from Pennsylvania State University, all in mathematics. Before forming Technology Strategies & Alliances, he was an executive vice president of Hambrecht & Quist Inc. (H&Q), an investment banking firm specializing in high-technology companies. Before joining H&Q, he was undersecretary of defense for research and engineering at the U.S. Department of Defense. He was one of the founders of ESL, Inc. and served as its president until he entered government service. Before that, he was with Sylvania/General Telephone and was the director of their Electronic Defense Laboratories. He is a member of the National Academy of Engineering and a fellow of the American Academy of Arts and Sciences. He has received the Army's Outstanding Service Medal, the Department of Defense Distinguished Public Service Award two years in a row, and NASA's Distinguished Service Medal. Other awards he has received include the Medal of Achievement by the American Electronics Association and the Knight Commander's Cross awarded by the Federal Republic of Germany.

BRIAN H. ROWE is senior vice president of General Electric's Aircraft Engines Department, headquartered in Cincinnati, Ohio. Mr. Rowe received his B.S. degree in mechanical engineering with honors from Kings College, Durham University. Most of Mr. Rowe's career has been with GE. He spent most of his early years with the company in design, development, and engineering, where he designed various types of vertical-takeoff-and-landing (VTOL) equipment. Mr. Rowe was elected vice president and general manager of the Airline Programs Division, and subsequently became vice president and general manager of the Aircraft Engineering Division. He is a fellow of the Royal Aeronautical Society and the American Institute of Aeronautics and Astronautics, and he is a member of the National Academy of Engineering. He was awarded the Legion d'Honneur by the French Government. He was also awarded an honorary degree of doctor in science and technology from the University of Cincinnati.

ALBERTUS D. WELLIVER is corporate senior vice president of the Engineering and Technology Division at the Boeing Company in Seattle, Washington. He earned a B.S. degree from Pennsylvania State University in mechanical engineering and completed the Stanford University Executive Business Program. After graduation from Penn State, Mr. Welliver joined the Research Division of Curtiss-Wright Corporation and remained there until he began his career with the Boeing Company. In his current position, he has responsibility for the company's critical, high-level engineering and technology development activities. He is a member of many professional organizations and has been selected as a fellow in the American Institute of

Aeronautics and Astronautics. He is a member of the NASA Aeronautical Advisory Board and the U.S. Air Force Scientific Advisory Board. He is also a member of the National Academy of Engineering, a fellow in the Royal Aeronautical Society, and a past chairman of the National Research Council's Aeronautical and Space Engineering Board. He has been honored as a Penn State Outstanding Engineering Alumnus.

Symposium Program

Symposium on the Future of Aerospace

February 28, 1992

1:00 p.m.	**Opening Remarks** Robert M. White, President National Academy of Engineering
1:05 p.m.	**Symposium Goals** H. Guyford Stever (symposium chairman) Commissioner, Carnegie Commission on Science, Technology, and Government
1:10 p.m.	**The Future of Manned Spaceflight** Aaron Cohen, Director NASA Johnson Space Center
	The Future of Aircraft Propulsion Brian H. Rowe, Senior Vice President GE Aircraft Engines
3:15 p.m.	**The Future of Large Passenger Aircraft** Albertus D. Welliver Corporate Senior Vice President Engineering and Technology The Boeing Company

The Future of Rotary-Wing Aircraft
Richard M. Carlson, Chief, Advanced Systems Research and Analysis Office,
U.S. Army Aviation Systems Command
NASA Ames Research Center

Defense Aerospace for a New World Order
William J. Perry, Chairman and
Chief Executive Officer
Technology Strategies and Alliances

Closing Remarks
H. Guyford Stever (symposium chairman)

5:30 p.m. **Reception**

Symposium Participants

Robert S. Aikenhead, Consultant, American Society of Mechanical Engineers, Irvine, California

Lew Allen, Jr., Chairman, The Charles Stark Draper Laboratory, Inc., Cambridge, Massachusetts

Jean Anderson, Aeronautics Librarian, California Institute of Technology, Pasadena, California

Lloyd Appelman, President, Aerospace Electrical Society, Anaheim, California

Henry and Edith Artof, Tustin, California

Holt Ashley, Professor Emeritus, Stanford University, Departments of Aeronautics/Astronautics and Mechanical Engineering, Stanford, California

J. Leland Atwood, (Retired President and Chief Executive Officer, Rockwell International Corp.), Pacific Palisades, California

William F. Ballhaus, Sr., President, International Numatics, Inc., Beverly Hills, California

Robert Ray Beebe, (Retired Senior Vice President, Homestake Mining Company), Consultant, Mendocino, California

Cristina Billingham, Research Assistant, Graduate School of Management, University of California, Irvine

William Boyer, Space News, Anaheim, California

Donald A. Brand, Senior Vice President and General Manager,

Engineering and Construction Business Unit, Pacific Gas and Electric Company, San Francisco, California

Barry Brayer, Manager, Quality and Environmental Staff, Federal Aviation Administration, Los Angeles, California

Marc H. Brodsky, IEEE Technology Admin. Fellow, Office of the Under Secretary for Technology, U.S. Department of Commerce, Washington, D.C.

Robert Bromberg, (Retired Vice President, Research, and Engineering, TRW Electronics, and Defense), Consultant, Los Angeles, California

George Bugliarello, President, Polytechnic University, Brooklyn, New York

John Burton, Rohr Industries, Inc., Chula Vista, California

Robert P. Caren, Vice President, Science and Engineering, Lockheed Corporation, Calabasas, California

Richard M. Carlson, Chief, Advanced Systems Research and Analysis Office (ASRAO), U.S. Army Aviation Systems Command, NASA Ames Research Center, Moffett Field, California

John W. Chambers, Sr. Vice President, Marketing and Sales, The Hartwell Corporation, Placentia, California

Hsien K. Cheng, Professor, University of Southern California, Department of Aerospace Engineering, Los Angeles, California

Richard M. Christensen, Senior Scientist, Lawrence Livermore National Laboratory, Materials Division, Chemistry & Materials Science Dept., Livermore, California

Leslie J. Cohen, McDonnell Douglas Astronautics Corporation, Huntington Beach, California

Aaron Cohen, Director, NASA Johnson Space Center, Houston, Texas

Frank E. Cole, President, BFM Transport Dynamics Corporation, Santa Ana, California

W. Dale Compton, Lillian M. Gilbreth Distinguished Professor of Industrial Engineering, Purdue University, School of Industrial Engineering, West Lafayette, Indiana

Thomas B. Cook, Jr., (Retired Executive Vice President, Sandia National Laboratories), Pleasanton, California

John R. Cook, Assistant Professor, Department of Industrial Engineering and Management, North Dakota State University, Fargo, North Dakota

George E. Cooper, Consultant, Saratoga, California

George Coryell, Engineering Specialists-Oper., Strategic Planning, General Dynamics/Convair, San Diego, California

Robert C. Crooke, (Retired President, Global Marine Development Inc.), Templeton, California
Jose B. Cruz, Jr., Professor, University of California, Department of Electrical and Computer Engineering, Irvine, California
W. Kenneth Davis, (Retired Vice President, Bechtel, Power Corporation and Former Deputy, Secretary of Energy), Consultant, Management and Engineering, San Rafael, California
Frank W. Davis, (Retired President, Convair, Aerospace Division and Fort Worth, Division, General Dynamics, Corporation), La Jolla, California
Gerald P. Dinneen, Foreign Secretary, National Academy of Engineering, Washington, D.C.
Allen F. Donovan, (Retired Senior Vice President, The Aerospace Corporation), Consultant, Corona del Mar, California
Carl Ehrlich, Rockwell International, Downey, California
Lloyd E. Elkins, Sr., (Retired Production Research, Director, Amoco Production Company), Petroleum Consultant, Tulsa, Oklahoma
James C. Elms, (Former Advisor to the Director, Strategic Defense Initiative Organization, and to Administrator of NASA), Newport Beach, California
Gerard W. Elverum, Jr., (Retired Vice President and General, Manager, Applied Technology, Division, TRW Space and Defense), Banning, California
Jack L. Ferrell, Vice President (retired), Manufacturing and Systems, TRW, Incorporated, Palos Verdes, California
Alexander H. Flax, Senior Fellow, National Academy of Engineering, Washington, D.C.
Richard G. Folsom, (President Emeritus, Rensselaer Polytechnic Institute), Napa, California
Irv Freund, Vice President of Marketing, British Petroleum (Hitco) Inc., Santa Ana, California
Lynne Friedmann, Consultant, Friedmann Communications, Solana Beach, California
Welko E. Gasich, (Retired Executive Vice President, Programs, Northrop Corporation), Encino, California
Ivan A. Getting, (President Emeritus, The Aerospace Corporation), Independent Consultant, Los Angeles, California
Norman A. Gjostein, Director, Materials Research Laboratory, Ford Motor Company, Dearborn, Michigan
George J. Gleghorn, (Retired Vice President, and Chief Engineer, TRW, Space and Technology Group), Rancho Palos Verdes, California

Mary L. Good, Senior Vice President - Technology, Allied-Signal Inc., Morristown, New Jersey
Lee Gordon, Client Service Officer, International Resource Center, San Diego, California
Peter Grefe, President, Shur-Lok Corporation, Irvine, California
Juergen Habermeier, President, Composite Materials Division, CIBA-GEIGY Corporation, Anaheim, California
Donald L. Hammond, (Retired Director, Hewlett-Packard Labs, Hewlett-Packard Company), Los Altos Hills, California
Robert C. Hansen, President, R. C. Hansen Inc., Tarzana, California
Willis M. Hawkins, Senior Advisor, Lockheed Corporation, Calabasas, California
Harold and Jane Hirsch, Pacific Palisades, California
William G. Howard, Jr., Independent Consultant, Scottsdale, Arizona
Chieh-Su Hsu, Professor Emeritus of Applied Mechanics, University of California, Department of Mechanical Engineering, Berkeley, California
Robert L. Johnson, (Retired Corporate Vice President, McDonnell Douglas Corporation), Laguna Niguel, California
Roy G. Johnston, Vice President, Brandow and Johnston Associates Consulting Structural Engineers, Los Angeles, California
Thomas V. Jones, (Former Chairman, Northrop Corporation), Los Angeles, California
Edward R. Kane, Former President, E.I. du Pont de Nemours & Company, Concord Plaza, Quillen Building, Wilmington, Deleware
Melvin F. Kanninen, Program Director, Engineering Mechanics, Southwest Research Institute, San Antonio, Texas
John R. Kiely, Consultant, Woodside, California
C. Judson King, Provost, Professional Schools and Colleges, and Professor of Chemical Engineering, University of California, Department of Chemical Engineering, Berkeley, California
Warren Knudson, Vice President, Sparta, Inc., Laguna Hills, California
Robert H. Korkegi, Consultant, Washington, D.C.
James N. Krebs, (Retired Vice President, General Electric Company), Santa Fe, New Mexico
Leslie Lackman, Vice President, Advanced Aircraft and Research Engineering, Rockwell International Corporation, North American Aircraft, El Segundo, California
George C. Larson, Editor, Air & Space/Smithsonian, Washington, D.C.

Jerome F. Lederer, (President Emeritus, Flight Safety Foundation; Retired Safety, Director, NASA; and Adjunct, Professor, Institute of Safety and System Management, USC), Laguna Hills, California
William E. Leonhard, Retired Chairman, President and Chief Executive Officer, The Parsons Corporation, Pasadena, California
Robert H. Liebeck, Fellow, McDonnell Douglas Corporation, Long Beach, California
Harold Liebowitz, L. Stanley Crane Professor of Engineering and Applied Science, The George Washington University, Washington, D.C.
T. H. Lin, Professor Emeritus, University of California, Department of Civil Engineering, Los Angeles, California
Walter B. Loewenstein, Energy Technology Consultant, Palo Alto, California
Paul A. Lord, California Polytechnic, Pomona, California
Artur Mager, (Retired Group Vice President, The Aerospace Corporation), Consultant, Los Angeles, California
John L. Mason, 1990 President, SAE International, Warrendale, Pennsylvania
John L. Mason, Consultant, Allied-Signal Aerospace Company, Palos Verdes Estates, California
Hudson Matlock, Consultant, Kerrville, Texas
Adolf D. May, Professor Emeritus of Civil Engineering, University of California, Civil Engineering Department, Berkeley, California
Bill B. May, Consultant, ARGO Systems, Inc., Sunnyvale, California
John L. McLucas, Aerospace Consultant, Alexandria, Virginia
Duane T. McRuer, President and Technical Director, Systems Technology, Inc., Hawthorne, California
Stephen A. Merrill, Executive Director, Office of Government and External Affairs, National Research Council, Washington, D.C.
Darrell Meyer, Vice President/General Manager, Elsinore Aerospace Systems, Chatsworth, California
Richard M. Morrow, Retired Chairman, Amoco Corporation, Chicago, Illinois
James H. Mulligan, Jr., Professor of Electrical Engineering, University of California, Irvine, California
Dale D. Myers, President, Dale Myers and Associates, Leucadia, California
Venkatesh Narayanamurti, Dean of Engineering, University of California, Santa Barbara, California

Nagy Nosseir, Professor, Department of Aerospace Engineering and Engineering Mechanics, San Diego State University, San Diego, California

Russell R. O'Neill, Dean and Professor of Engineering Emeritus, University of California, School of Engineering and Applied Science, Los Angeles, California

Paul Olk, Graduate School of Management, University of Califorina, Irvine

Herschel G. Owens, Director, Engineering Systems, Rockwell International, Seal Beach, California

Norman F. Parker, (Retired President and Chief Executive Officer, Varian Associates), Cardiff by the Sea, California

Robert J. Parks, (Retired Deputy Director, Jet Propulsion Laboratory), Balboa Island, California

Joseph A. Pask, Professor Emeritus of Ceramic Science and Engineering, University of California, Department of Materials Science and Mineral Engineering, Berkeley, California

C. Kumar N. Patel, Executive Director, Research Materials Science, Engineering and Academic Affairs Division, AT&T Bell Laboratories, Murray Hill, New Jersey

George Paulikas, Group Vice President, Programs, Aerospace Corporation, El Segundo, California

David and Anne Pearsall, Raleigh, North Carolina

Jerry Perazzo, GE Aircraft Engines, General Electric Company

Courtland D. Perkins, (Past President, National Academy of Engineering), Consultant, Alexandria, Virginia

William J. Perry, Chairman and Chief Executive Officer, Technology Strategies & Alliances, Menlo Park, California

William H. Pickering, President, Lignetics Inc., La Canada, California

John R. Pierce, Professor, Stanford University, Center for Computer Research in Music and Acoustics, Stanford, California

Allen Plotkin, Department of Aerospace Engineering and Engineering Mechanics, San Diego State University, San Diego, California

William R. Prindle, (Retired Vice President and, Associate Director-Technology, Corning Incorporated), Consultant, Santa Barbara, California

Allen E. Puckett, (Chairman Emeritus, Hughes Aircraft Company), Pacific Palisades, California

Abner and Henrietta Rasumoff, Rolling Hills Estates, California

Eberhardt Rechtin, Professor of Engineering, University of Southern California, Los Angeles, California

Larry G. Redekopp, Professor and Co-Chairman, Department of Aerospace Engineering, University of Southern California, Los Angeles, California

John Reinert, Project Manager, Elsinore Aerospace Systems, Chatsworth, California

Dennis K. Rice, Vice President - Planning, Northrop Corporation, Los Angeles, California

Ron Richmond, Director of Programs, Brunswick Defense, Costa Mesa, California

Louis W. Riggs, (Retired Chairman, Tudor Engineering Company), Lafayette, California

Fanny Rivera, Deputy Regional Administrator, Federal Aviation Administration, Los Angeles, California

Robert K. Roney, (Retired Senior Vice President, Hughes Aircraft Company), Santa Monica, California

Leonard G. Rosenbaum, District Manager-Southern California, Business Planning and Market Development Operation, GE Aircraft Engines, General Electric Company, Long Beach, California

Brian H. Rowe, Senior Vice President, GE Aircraft Engines, Cincinnati, Ohio

Warren G. Schlinger, (Retired Laboratory Manager, and Director, Texaco, Inc.), Consultant, Pasadena, California

David K. Schmidt, Director, Aerospace Research Center, Arizona State University, Tempe, Arizona

Arthur Schnitt, Consultant, Los Angeles, California

Manfred R. Schroeder, Direktor, Drittes Physikalisches Institut, University of Goettingen, Germany

Harris M. Schurmeier, (Retired Associate Director, Jet Propulsion Laboratory), Fallbrook, California

Jonathan K. Scudder, Chief Scientist, Litton Industries, Inc., Beverly Hills, California

F. Stan Settles, Consultant, Tempe, Arizona

Bill Shell, Chairman Emeritus, California Engineering Foundation, Northrup University, Brea, California

Professor Shan-Fu Shen, John Edson Sweet Professor of Engineering, Cornell University, Sibley School of Mechanical and Aerospace Engineering, Ithaca, New York

Ernest T. Smerdon, Dean of Engineering and Mines, University of Arizona, Tucson, Arizona

A.M.O. Smith, Retired Aerodynamics Engineer, San Marino, California

Leroy H. Smith, Jr., Manager, Turbomachinery Aero Technology, GE Aircraft Engines, Cincinnati, Ohio

Dean R. Snipes, Consultant and Manufacturing Manager, Anaheim, California
Joan S. Spencer
William J. Spencer, President and Chief Executive Officer, Sematech Inc., Austin, Texas
Zoltan Z. Stacho, President, Holmes & Narver, Inc., Orange, California
Sam Stameson, Group Vice President, Engineering and Technology, Hughes Aircraft Company, Fullerton, California
Harry Staubs, Senior Staff Specialist, McDonnell Douglas Space Systems Company, Seal Beach, California
Richard J. Stegemeier, Chairman and Chief Executive Officer, Unocal Corporation, Los Angeles, California
Morris A. Steinberg, (Retired Vice President, Science, Lockheed Corporation), Consultant, Los Angeles, California
William D. Stevens, (Retired Chairman, Foster Wheeler Corporation), Dennis, Massachusetts, North Caldwell, New Jersey
H. Guyford Stever, Commissioner, Carnegie Commission on Science, Technology, and Government, Washington, D.C.
Allen R. Stubberud, Department of Electrical and Computer Engineering, University of California, Irvine, California
Ivan E. Sutherland, Vice President and Sun Fellow, Sun Microsystems Laboratories Inc., Mountain View, California
John M. Swihart, President, National Center for Advanced Technology, Washington, D.C.
Clarence A. Syvertson, (Retired Director, NASA Ames Research Center), Saratoga, California
Dean Takahashi, Staff Writer, Orange County Edition, Los Angeles Times, Costa Mesa, California
Morris Tanenbaum, (Retired Vice Chairman and Chief Financial Officer, AT&T), Short Hills, New Jersey
Joseph B. Tarlton, Manager of Space, Bechtel National Inc., San Francisco, California
Judith A. Thams, Development Planning Specialist, Northrop Corporation, Aircraft Division, Hawthorne, California
Mitchell Thomas, President and Chief Executive Officer, L'Garde Inc., Tustin, California
Donald O. Thompson, Professor of Aerospace Engineering and Engineering Mechanics, and Director, Center for NDE, Iowa State University, Ames, Iowa
Caroline L. Vaughan, Director, Newmarket Venture Capital, United Kingdom

Hank Verbais, Aviation Education Officer, Western Pacific Region, Federal Aviation Administration, Los Angeles, California
Walter K. Victor, Senior Associate, Technical Arts Associates, Pasadena, California
Irving T. Waaland, Vice President and Chief Designer, Northrop Corporation, Pico Rivera, California
Larry D. Welch, President, Institute for Defense Analyses, Alexandria, Virginia
Albertus D. Welliver, Corporate Senior Vice President Engineering and Technology, The Boeing Company, Seattle, Washington
Elmer P. Wheaton, Director and Associate, Marine Development Associates, Inc., Portola Valley, California
Harold A. Wheeler, (Retired Consultant, Hazeltine Corporation), Ventura, California
Albert D. Wheelon, (Retired Chairman and Chief Executive Officer, Hughes Aircraft Company), Los Angeles, California
Robert M. White, President, National Academy of Engineering, Washington, D.C.
J. Ernest Wilkins, Jr., Distinguished Professor of Applied Mathematics and Mathematical Physics, Clark Atlanta University, Atlanta, Georgia
Basil W. Wilson, (Retired Private Consulting Oceanographic Engineer), Pasadena, California
Bertram Wolfe, Vice President and General Manager, GE Nuclear Energy, San Jose, California
Arch Wood, Executive Director, Commission on Engineering and Technical Systems, National Research Council, Washington, D.C.
Herbert H. Woodson, Dean of Engineering, The University of Texas, Austin, Texas
R. Barry Wrenn, President, British Petroleum (Hitco) Inc., Santa Ana, California
Theodore Y. Wu, Professor of Engineering Science, California Institute of Technology, Pasadena, California
Henry T.Y. Yang, Dean, Schools of Engineering and Neil A. Armstrong Distinguished Professor, Purdue University, West Lafayette, Indiana
Abe M. Zarem, (Retired Founder, President, and, CEO, Electro-Optical Systems, Inc., and Xerox Development Corp., A Subsidiary of Xerox Corporation), Beverly Hills, California
Katherine Zin, Research Assistant, Graduate School of Management, University of California, Irvine
John Zuk, Chief, Aviation Technology Office, NASA, Ames Research Center, Moffett Field, California
Maureen Zuk